Power Systems

C. Rehtanz
Autonomous Systems and Intelligent Agents
in Power System Control and Operation

Springer

Berlin
Heidelberg
New York
Hong Kong
London
Milan
Paris
Tokyo

Engineering ONLINE LIBRARY
http://www.springer.de/engine/

C. Rehtanz

Autonomous Systems and Intelligent Agents in Power System Control and Operation

With 153 Figures

 Springer

PD Dr.-Ing. Christian Rehtanz
ABB Switzerland Ltd.
Corporate Research
5405 Baden-Dättwil
Switzerland

Cataloging-in-Publication Data applied for
Bibliographic information published by Die Deutsche Bibliothek
Die Deutsche Bibliothek lists this publication in the Deutsche Nationalbibliografie;
detailed bibliographic data is available in the Internet at <http://dnb.de>

ISBN 3-540-40202-0 Springer-Verlag Berlin Heidelberg New York

Springer-Verlag Berlin Heidelberg New York
a member of BertelsmannSpringer Science + Business Media GmbH

http://www.springer.de

© Springer-Verlag Berlin Heidelberg 2003
Printed in Germany

Typesetting: Digital data supplied by authors
Cover-Design: de'blik, Berlin
Printed on acid-free paper   62/3020 Rw 5 4 3 2 1 0

Preface

Autonomous systems are seen from the scientific and industrial perspective as one of the most important trends for the next generation of control systems.

The topic of autonomous systems covers the modularization of software applications and their special organization to achieve an intelligent and autonomous behavior. These modules can be seen as agents. The computer science has designed algorithms for agent interactions from a theoretical point of view. Agent technologies belong to the area of computational intelligence, because they enable systems with a certain degree of intelligent behavior. Intelligence in this context mainly means less user interactions and automated adaptation to changing environments. Intelligence reflects the transformation of low value data into higher value information. In this case, the knowledge is not produced by itself, but existing knowledge is made usable. By introducing computational intelligence and agent technologies, human interactions with the system are reduced and concentrated to most relevant areas, which means that a higher degree of autonomous behavior is achieved.

In the last years, several books about computational intelligence applications in power systems have been published. Agent technologies are the latest developments and are now on the edge between theory and application for power systems, but they are not covered in existing books for power systems yet. This book is the first one, which transfers autonomous system and agent's theory into the power system's control and operation environment. This new approach is applied to several applications for power systems.

Motivation

Why autonomous systems are such an important trend in power system's control, can be derived as follows. The change of the paradigms in the electrical power industry leads to changing responsibilities of the market players. The management of the market activities changes the areas of economics and technology drastically. The billing is only one example for a strongly changed area. Only the introduction of computerized and automated billing systems enabled the implementation of wide spreading market activities down to every single private household. From contract handling to supply organization, all the functions are performed electronically and automatically hand in hand. Just the development of the Information Technology (IT) and data processing have built the base for market and trading activities with the desired extent. The development of IT is strongly coupled with the development of the energy sector. This enforces developments on both sides.

The changes in both process and business level structures in utilities impose the requirements of matching and linking information of the market activities and technical process to fulfill the constraints on both sides simultaneously. Due to the unbundling of the particular areas, less information is available directly. Therefore an exchange has to be established explicitly between market participants and network operators. This data exchange has to be provided by IT solutions. Due to all these changes of the process environment, the requirements for the control and operation systems are altering, as well.

Considering a more detailed view of these changes, the power system operation has to react and be adapted flexibly to more volatile operational situations. The operation planning has to take the market behavior into account. The optimization of the network operation for instance has to consider modified constraints due to the limited access to the power generation operation. The operation of the generation units is as well no longer determined only by technical and enterprise internal economical constraints, but also by market strategies. The network itself does not have to provide a maximum reliability, but a maximum return on investment for the network owners, in which the reliability is only one out of many factors to consider. On the other hand this influences asset management strategies like maintenance and network planning.

Conclusively, the complexity as well as the volatility of the operational tasks will increase. This requires a fundamental increase of the degree of automation, because interactions of the process operator are constrained by speed and complexity limits. To fulfill these new requirements it is essentially necessary further develop decision support and automation by using and pushing recent trends of agent and IT technologies.

Focus and Target

The target of this book is to design and apply a control system architecture, which provides and uses data and information on the different process levels appropriately. The various complex and volatile processes have to be modeled and strategies for their management have to be developed. Knowledge on how to operate such a process has to be implemented in an automated form. Procedures to make knowledge usable in an automated way can be summarized under the acronym computational intelligence.

The focus of this book is to design a future control system architecture for electrical power systems, which copes with the changed requirements concerning complexity and flexibility. The progressing IT has to be exploited to enable intelligent control systems.

The concept of autonomous systems, used for electrical power control systems is developed and applied exemplarily to fulfill all these changed requirements and to enable new capabilities. It builds the structural background for future developments in control and operation systems. Essentially, components are designed to take over parts of the automation task, which are interacting with each other. Information provided by IT and communication systems will be processed and autonomous actions are taken on pure pre-defined criteria. In this case knowledge has to be modeled, stored and processed to enable the autonomous actions.

This book is focusing both on the concept of autonomous systems and on its application for the implementation of intelligent techniques. This autonomous architecture shall enable especially new solutions based on intelligent agent techniques, which are providing capabilities to fulfill the above-defined requirements. To clarify the applicability and profitability of the new approach, several applications for different particular tasks in power system operation and control are presented in detail.

The target audiences of this book are electrical, control and information engineers in science, industry and utilities with basic background in power systems. It is assumed that the basic questions and procedures in power system control and operation are known. The content of the book shall be helpful and inspiring for people responsible for design, engineering, research and development of control and operation systems for power systems. Beside the future structure of control systems, this book informs also about agent technologies and their application to power systems.

Structure

The first part of this book builds the theory of autonomous systems which is adapted to power systems. The autonomous system architecture provides the capabilities to implement intelligent agent technologies. Chapter 1 presents this new approach for the logical structure of control systems for electrical power systems. As a step further from the basic concept, chapter 2 discusses the autonomous system structure from the information technology point of view. Aspects of communication, standard data modeling and IT security issues are included here to prove the practical implementation of this approach. The value of this approach lies in an open system architecture, which enables flexible and extendable control systems as well as decentralized automatically acting systems. The set up of a higher degree of decentralization results in the use of computational intelligence and intelligent agents, which are introduced in chapter 3. Chapter 4 presents details of negotiation theory as a base for power system applications.

The second part of the book, starting with chapter 4, presents applications of the autonomous system architecture. The main focus lies on intelligent agent applications showing solutions profitable in several particular areas of power system control and operation. All these topics show both the scientific novelty and the practical applicability. After the thorough review of the agent negotiation theory, chapter 4 applies this to energy market negotiations and maintenance scheduling issues. Chapter 5 deals with disturbance diagnosis. Chapter 6 proposes a power system restoration scheme. Chapter 7 applies agents to protection systems. Chapter 8 presents applications to damping control, chapter 9 to coordinated FACTS control and chapter 10 to secondary voltage control. Chapter 11 proposes control center applications, especially visualization schemes. Chapter 12 contains applications for controlled islanding and microgrid operations. Chapter 13 proposes a quality control center approach. All these application chapters are covering a broad field of innovative and promising approaches for the usage of the proposed autonomous system architecture and intelligent agents.

In conclusion, this book draws the circle from the theoretical and IT-concept of autonomous systems for power system control over the required knowledge-based methods, especially intelligent agent techniques, and their capabilities to concrete applications within this field.

Acknowledgements

I would like to thank all the co-authors providing excellent research and development results to this book. The challenging task of writing and editing this book was possible by the well-prepared contributions of a team of international researchers believing in the need of the presented contents.

My sincere thanks to all contributors, proofreaders, the publisher and my family for making this book project happen.

Christian Rehtanz   Baden-Daettwil, Switzerland, April 2003

Contents

Contributors

Becker, Christian, Dr.
Institute of Electric Power Systems, University of Dortmund
Emil-Figge-Str. 20, 44221 Dortmund

since 2002
AIRBUS Deutschland GmbH
Environmental Control, Dept. EYVCC
Kreetslag 10, 21129 Hamburg
Germany

Coury, Denis V., Prof
Department of Electrical Engineering,
School of Engineering of São Carlos, University of São Paulo
São Carlos, SP 13566-590
Brazil

Giovanini, Renan
Department of Electrical Engineering,
School of Engineering of São Carlos, University of São Paulo
São Carlos, SP 13566-590
Brazil

Hasegawa, Jun, Prof.
Division of System and Information Engineering
Graduate School of Engineering, Hokkaido University
Sapporo
Japan

Honavar, Vasant, Prof.
Electrical and Computer Eng.,
Iowa State University
USA

Hopkinson, Kenneth M.
Department of Computer Science, Cornell University
Ithaca, NY 14853
USA

Hossack, John A.
  Institute for Energy and Environment
  Department of Electronic & Electrical Engineering, University of Strathclyde
  204 George Street, Glasgow G1 1XW, Scotland
  United Kingdom

Kita, Hiroyuki, Prof.
  Division of System and Information Engineering
  Graduate School of Engineering, Hokkaido University
  Sapporo
  Japan

Kostic, Tatjana, Dr.
  ABB Switzerland Ltd - Corporate Research
  Segelhof, CH-5405 Baden-Daettwil
  Switzerland

Leder, Carsten, Dr.
  Institute of Electric Power Systems, University of Dortmund
  Emil-Figge-Str. 20, 44221 Dortmund
  Germany

Lefebvre, Serge, Dr.
  Institut de Recherche d'Hydro-Québec (IREQ)
  1800 Blvd Lionel Boulet, Varennes, Quebec, J3X 1S1
  Canada

Li, Hao
  Department of Electrical Engineering, University of Washington
  Seattle, WA 98195
  USA

Liu, Chen-Ching, Prof.
  Department of Electrical Engineering, University of Washington
  Seattle, WA 98195
  USA

McArthur, Stephen D.J., Dr.
  Institute for Energy and Environment
  Department of Electronic & Electrical Engineering, University of Strathclyde
  204 George Street, Glasgow G1 1XW, Scotland
  United Kingdom

McCalley, James D., Prof.
Electrical and Computer Eng.
Iowa State University
USA

McDonald, James R., Prof.
Institute for Energy and Environment
Department of Electronic & Electrical Engineering, University of Strathclyde
204 George Street, Glasgow G1 1XW, Scotland
United Kingdom

Naedele, Martin, Dr.
ABB Switzerland Ltd - Corporate Research
Segelhof, CH-5405 Baden-Daettwil
Switzerland

Nagata, Takeshi, Prof.
Department of Information and Intellectual Systems Engineering
Faculty of Engineering, Hiroshima Institute of Technology
2-1-1 Miyake, Saeki-ku, Hiroshima 731-5193
Japan

Rehtanz, Christian, Dr.
ABB Switzerland Ltd - Corporate Research
Segelhof, CH-5405 Baden-Daettwil
Switzerland

Sasaki, Hiroshi, Prof.
Department of Electrical Engineering, Hiroshima University
1-4-1, Kagamiyama, Higashi-Hiroshima City, 739-8527,
Japan

Thorp, James S., Prof.
School of Electrical Engineering, Cornell University
Ithaca, NY 14853
USA

Vishwanathan, Vijayanand
Open Systems International, Inc.
3600 Holly Lane N Suite # 40, Minneapolis, MN 55447
USA

Haifeng, Wang, Dr.
   Department of Electronics and Electrical Engineering, University of Bath
   Bath, BA2 7AY
   United Kingdom

Wang, Xiaoru, Dr.
   School of Electrical Engineering, Cornell University
   Ithaca, NY 14853
   USA

Zhang, Zhong
   Electrical and Computer Eng.
   Iowa State University
   USA

Zoka, Yoshifumi, Dr.
   Department of Electrical Engineering, University of Washington
   Seattle, WA 98195
   USA

1 Autonomous Control System Architecture

Christian Rehtanz

ABB Switzerland Ltd - Corporate Research, Baden-Daettwil, Switzerland

This chapter introduces the basic architecture of autonomous systems for intelligent control and operation systems for use in power systems. The functional architecture enabling the application of agent techniques is based on fundamental theoretical aspects. The operational architecture for control systems of power systems is defined and characterized. The proposed autonomous system structure enables the thorough usage of Information Technology (IT) to achieve a higher degree of automation and a more reliable operation of electrical power systems.

1.1 Fundamental Aspects of Autonomous Systems

An autonomous system is a system, which can react intelligently and flexibly on changing operating conditions and demands from the surrounding processes. The system works analogously to the object-oriented and component-based paradigms in Computer Science, in which a request or inquiry is directed to an object or a component, which is then expected to show a desired reaction. In the autonomous system, the inquiry is directed to an autonomous component that acts, or answers appropriately either in form of an activity or with information. The inquiry is processed more or less autonomously, depending on the process status and the boundary conditions. Autonomous in this sense means that only the target of the inquiry is set, but the way to the solution is chosen in a self-responsible manner. With increasing responsibility for the actions, the extent of autonomy grows as well. Because of this, additional knowledge has to be integrated into the autonomous components. Therefore, they can be labeled in a certain sense as intelligent.

The meaning of autonomous acting can be explained with an example from the robotics area. A robot on the moon is supposed to play with a moving ball. The robot is controlled from the earth. The conventional process control would use the sensors of the robot to detect the ball and send the sensor-signals to the controller on earth. The controller would define moving actions for the robot and transmit these as signals back to the robot on the moon, which is executing these movements according to the signals from earth. Due to the long distances, the delay times for the signals are so long that the actions are not fitting to the actual situation, which has changed in between.

Autonomous control in this case would mean, that the robot has predefined knowledge about its possible moving actions, e.g. the action 'catch the ball'. The controller on the earth would generate and transmit only high value action settings like 'catch the ball'. This means that the controller organizes or manages the actions without defining all the movement steps. The robot itself is translating these high value action settings into concrete action steps fitting to the actual local situation, like the actual position of the ball. These action steps are for example 'go forward' or 'lift the arm'. The robot would coordinate its movements. Likewise he would adapt his actions according to barriers or even other robots. The actions of the generated movements are executed in the last step. The entire procedure distinguishes between Organization/Management, Coordination and Execution. The high value action setting remains to the operator, whereas the low value or simple action steps are defined and taken by the robot as a local component. Therefore the intelligence of the robot is increased and autonomous acting is enabled to a certain extend. The focus of the following investigation is to apply this kind of control organization and structure to the area of electrical power system control and operation.

The original autonomous systems approach comes from the area of robotics [1][2]. The simplest example for this is a robot that is able to find a possible and favorable way under the specification of a target position independently. An extension on traffic-routing systems is described among other applications in [3]. Single power system applications, which are targeting in the direction of autonomous actions, are introduced in [4],[5] and [6]. The following descriptions of autonomous systems take up the aspects from [1] and [2], and propose the application for control and operation systems of power systems. First approaches of this area for control systems can be found in [7] and [8].

An autonomous system may consist of autonomous components, which represent likewise autonomous systems in the sense of subsystems. If the target definition and the resulting action of an autonomous system serve for controlling a process, this is called an autonomous control system. If it serves for the development of high order information, it is an autonomous information system. An example for this is the component for the supervision of a process or a system, with diagnostic information as result, which is made available to both the process operator and to other autonomous components. Beside the self-organization such a system allows also the self-diagnosis, which generates information about the accessibility of the given target.

The advantage that can be drawn from such architecture is that it acts over longer intervals and especially within broader boundaries, without having to intervene manually into the system. Especially in extraordinary or fast changing situations, the system acts adequately without intervention from outside. Autonomous systems are making sense in complex processes, which are underlying high and flexible demands. This is closely related to the fact that actions must be taken despite of little a-priori knowledge and uncertainty.

This uncertain knowledge is closely connected to the field of artificial and computational intelligence [9]. Intelligent abilities, such as learning from experience, planning of actions or detection and identification of errors, used to be dealt with by the process operator, are integrated more and more into the autonomous system. The advantages are reflected in the avoidance of operating errors, the enhancement of the reaction rate, and the relief of the operator [10].

1.2 Functional Architecture of Autonomous Components

The functional architecture of an autonomous component can be specified at first according to Fig. 1.1 independently of its application domain.

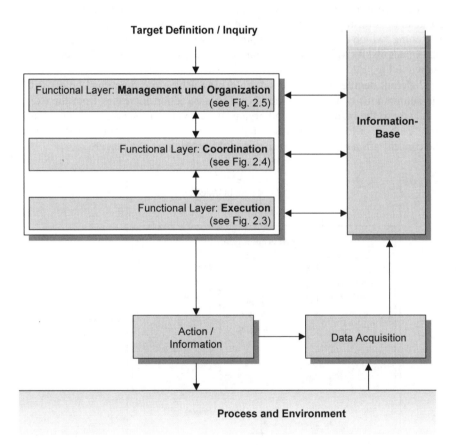

Fig. 1.1. Structure of an autonomous component

It contains the functional layers 'management and organization', 'coordination' and 'execution' [1]. The general structure of one functional layer is shown in Fig. 1.2. The detailed tasks are defined in section 1.2.1, Fig. 1.3 for the execution layer, in section 1.2.2, Fig. 1.4 for the coordination layer, and in section 1.2.3, Fig. 1.5 for the management and organization layer.

The autonomous component determines an action or information according to a target definition or inquiry. This could be for example the signal for keeping the power flow on a particular line or the question for a forecasted load value. The action or information is affecting the process or environment of the component, e.g. the power flow is kept or the forecasted value is available. The autonomous component acts self-responsible and independent as long as there are no changes in the target definition, also when the process or the environment is changing. The outgoing information can be used in hierarchical arrangements of several autonomous components as target definition for underlying components.

The three functional layers within the autonomous component consist of a unit for data or information connection or processing, a unit for logical conclusion or reasoning, which concludes the output from the data or information, and the output values itself, which are the inputs to control the next layer. This structure is shown in Fig. 1.2.

Different degrees of abstraction are distinguishing the functional layers within the autonomous component. The degree of abstraction increases from the execution layer to the management and organization layer. This separates different tasks from each other to reduce in each step the complexity of a problem and to enable the concentration of the particular part of the problem in each layer.

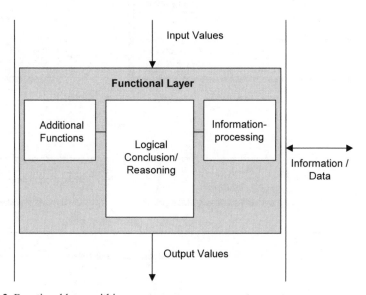

Fig. 1.2. Functional layer within an autonomous component

This idea is the same as the hierarchical layer structure of virtual machines in the information technology, in which the actions are translated from an abstract interpreter language via a high level code down to the machine code and executable micro instructions. For example keeping the power flow is a management task, which has to be coordinated with other network controllers under specific conditions like within normal or disturbed situations and finally executed.

The basis for these actions is a connection to an information base of the process and the environment, which provides all directly taken process data as well as high level information generated from other components. The particular layers of the autonomous components are getting data and information from this information base, but also providing information for other components. The following sections describe, which functions these layers have.

1.2.1 Execution Layer

To understand the functions of an autonomous component, the execution layer according to Fig. 1.3 is explained at first. A concrete action for the process results if a particular kind of action execution is defined.

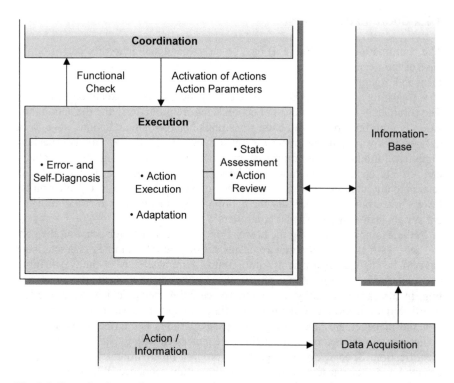

Fig. 1.3. Execution layer of an autonomous component

The response of the process is coupled back to the action execution via the data acquisition and the information base. A conventional control loop is defined if the action execution is a controller. The action execution can be adapted to the process through a continuous state assessment. Small continuous changes are considered here. An action review is also taking place within this block, which provides information about the function and quality of the actual execution of actions. A further part of the execution layer is the error- and self-diagnosis, which is similar to a physical check of the function regarding data connection and availability of the execution of actions. This information is transferred to the overlaid coordination level together with the action review information.

The execution layer, which is explained so far, is relatively limited in its actions. The addition of the coordination layer extends the action to the reaction on events, which are corresponding to major changes in the process. Together with this extension the execution layer has to contain different alternatives, which react on several particular process and environmental situations. Therefore the execution layer consists of multiple parallel layers regarding the action execution. In which sequence and with which starting parameters these particular action schemes are executed is determined by the coordination layer.

1.2.2 Coordination Layer

The coordination layer (Fig. 1.4) has the task to activate action execution on the base of the general progress definition and defined action steps from the management and organization layer.

It transmits also appropriate parameters for the particular execution of actions. The parameters for the sequence of action steps are predefined and the action transition is coordinated. These action transitions between single actions have to be defined in a way that the process is not influenced negatively. This is especially important for autonomous control systems, because stepwise controller switchings can activate unwanted excitations of the system. The structural change in the direction to autonomous control systems has no major effect on the controller on the lowest level, but adds especially the coordination functionality to react on different system states with particular controllers.

In the example of keeping the power flow there is a need for adapting the control settings if the system structure changes. For example damping characteristics have to be mapped permanently into control parameters. Also faulty situations may change the control settings drastically. Another example is the protection setting. In the case of a faulty protection device neighboring devices have to adapt their behavior according to the new situation. The general task from the organization and management layer to protect a particular area is unaffected by this. Particular sets of parameters for the actions are generated with the help of the process identification. Special events, e.g. major process or environmental changes, are detected and they cause modifications of the action sequence. Which event leads to which actions, is defined generally by the overlaid organization and management layer.

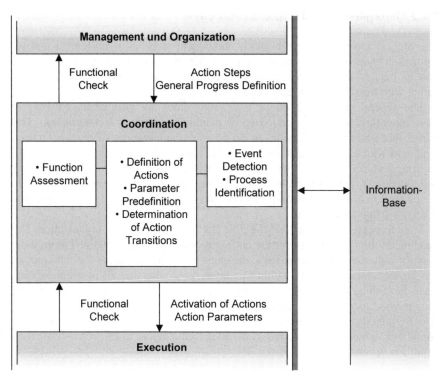

Fig. 1.4. Coordination layer of an autonomous component

The result of the functional check of the execution layer can be used for a direct parameter adaptation in the coordination layer. The detection of missing data, like measurement values or a disturbed data connection for instance, may lead to the immediate switching into an emergency operation state, like in the protection example above. Due to this the security of the process can be hold as long as possible without external intervention.

In addition, the coordination layer contains a function assessment. This adds to the functional check from the underlying execution layer assessments by the coordination layer. Information is delivered if appropriate actions for the actual situation can be activated or if the extension or change of the general action definition is necessary. It is supervised, if the existing functionality is sufficient or if the autonomous component is reaching its limit and needs to learn new action schemes. This is necessary for instance in the protection case when there is no knowledge available how to adapt the parameters of a protection device in the case of a faulty neighbor. The entire information of the function assessment is delivered to the highest layer of the autonomous component.

1.2.3 Management and Organization Layer

The highest layer of the autonomous component is called management and organization layer, which is shown in Fig. 1.5. This layer is getting the target definition or inquiry from the process operator or other autonomous components together with the related boundary conditions.

The general progress definition and action steps are planned on the base of these input targets, which is called action planning and action decomposition. This means that a function evaluation checks whether the target can be reached in principle and which action steps are necessary. Information is given back to the requesting instance if a particular target cannot be reached. In a control case the way is defined here, how to find an optimal controller within an automated process. In the protection example general protection and adaptation strategies are stored here in a knowledge-based form.

State forecasts help to extend the time horizon for the action organization. Depending on their importance for the entire process, they can be defined as an independent autonomous information component to which inquiries for the forecast can be sent. State forecasts within this layer mean also that all possible existing states have to be determined and appropriate action schemes have to be organized. This method extends the normal operation case for preventive predetermined actions. Optimal action schemes are generated for particular events or boundary conditions, which are transferred to the underlying layers to be prepared for their condition dependent activation.

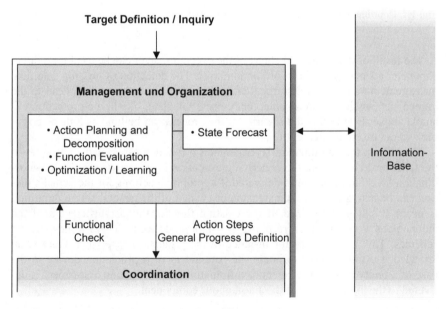

Fig. 1.5. Management and organization layer of an autonomous component

These could be for example control or protection parameters for different load, configuration or emergency resp. fault conditions. Due to this the time consuming tasks of planning are separated from the real time execution of actions. This preventive method shortens the response time after complex events or process changes.

An important function of this layer is learning. If the function evaluation together with the function assessment from the underlying layers has determined that an appropriate action is impossible, a new action scheme, which means new actions for a particular situation, have to be generated. For example, optimization methods have to find a controller with appropriate parameters for an autonomous control component within the search space of all possible controllers. This controller has to have an optimal characteristic for a new, up to this point not considered situation. The autonomous component can use this particular controller in the future for all similar situations and therefore it has learned and extended the capability of the autonomous component. This method goes further than just an adaptive control, which can be realized within the execution layer. Here a self-organizing and self-learning control is defined. In general the set of action schemes is expanded for particular situations and events.

The actions steps or progress are given as information to the underlying coordination layer, which coordinates their execution in the execution layer as described before. The generation of action steps contains a higher degree of abstraction than the functions in the lower levels. Therefore it can be stated as a generalized information processing.

The information of all the particular function checks is, beside the use within management and organization layer, attached to the information base. This self-supervision and diagnosis of the autonomous component is used for the determination of action modifications as well as in other systems, like maintenance systems for instance, or as hint for a necessary manual intervention. The limits of the autonomy are therefore transparent.

Both the management and organization layer, and the coordination layer are serving to take decisions. This is to be implemented as an abstract and mainly discrete process. The implementation of the functions within the execution layer is mostly a continuous or, due to a smoothing of the action transitions, quasi-continuous process. Therefore the autonomous component as a whole contains a hybrid continuous-discrete architecture.

1.2.4 Information Base

The information base has to be seen as a hybrid setup of direct data access to other components and a distributed database. Information and data are accessible via requests of the autonomous components through a data interface, which are represented in the figures above only as arrows. The locally used part of the data can be stored within the autonomous component to enable a fast access.

Three logical channels can represent the communication with the information base from the different layers of the autonomous components. The first channel contains the directly accessible process variables. These values have to be available in real-time for control and protection tasks.

Based on these data, the particular layers of the autonomous components generate high value process information, such as the state assessment, the process identification, the state forecast and the event detection. This calculated information can be made accessible to other components if necessary. The time ranges in which these information are available depends on the complexity and on the layer in which they are generated. The event detection, for example, has to be determined very fast. In comparison, the state forecast within the management layer is less time-critical. The information, which the information components provide as an output, belongs also to this channel of higher value process information. The kind and task of these components determine the time range in which this information is accessible. This could possibly be a boundary condition for the inquiry to these components.

The third channel are the combined diagnosis and function information, which are coordinating the autonomous components between each other, e.g. to implement reserve functions. They are also detecting the limits of autonomy and request manual interventions for extension or maintenance.

1.3 Structures of Autonomous Systems

Up to now, the description was focused on a single autonomous component. But a complex technical system with several tasks contains many autonomous components, which are building together an autonomous system. An autonomous component as a self-responsible acting unit can also be called agent, which is acting within its environment. If several of such agents are acting simultaneously, the whole system can be called a multi-agent system.

The literature describes systems for different fields of applications consisting of multiple agents or autonomous components. The relation of the components to each other distinguishes these systems. An autonomous system is more or less organized as a hierarchical structure. In comparison to that, multi-agent systems are often structures with a couple of similar agents within one level of hierarchy [4][5].

A hierarchical structure of autonomous subsystems or components, according to Fig. 1.6, enables a separation of particular functions and sub-functions on different levels of the structure. An overlying autonomous component can call a couple of underlying components. Such an overlying autonomous component is working on an abstract level. Its output information contains target definitions for the underlying components. Likewise process-covering information can be generated on a higher hierarchy level.

The implementation of the structure as a multi-agent system with similar agents on the same hierarchy level, as shown in Fig. 1.7, uses the structure and function of the autonomous component in a further way. The single components are trained in a way, that their behavior is following a global process target.

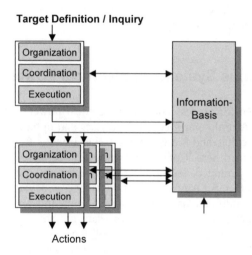

Fig. 1.6. Autonomous system with hierarchical structure

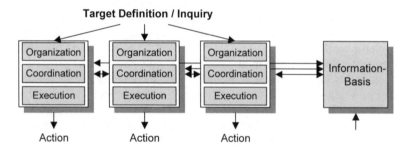

Fig. 1.7. Autonomous system with multi-agent structure

This means, that each component acts locally and as far as possible on local data, but despite of this the global target is reached in the sense of an optimal global behavior.

In addition to this capability, the communication between the components can be used to avoid counterproductive actions. Therefore the coordination layers of the autonomous components are transmitting action amplification or limiting information. Neighboring components can use this information in the case of a component failure and take over as far as possible the responsibility for the action of the failing component. This information is used from neighboring components if it is available. Otherwise, the components have to work completely autonomously. The same holds for general process information in hierarchical systems.

In the area of electrical power systems, with a process organized in a multiple parallel but also hierarchical way, both structures have to be applied. The following section explains how this functional architecture and structure can be translated and applied to the control and operation system of electrical power systems together with the benefits of its application.

1.4 Autonomous Systems in Electrical Power Systems

The control system for process control of electrical power systems is structured hierarchically. A central control center supervises and optimizes the process. Operations like switching and slow control tasks, usually triggered by a human operator, are executed here. Control and protection actions, which need a faster reaction time, are arranged decentralized. These actions are taken mostly with fixed sets of parameters. Any coordination is operated via the central control center as well as modifications of the criteria and parameters for the control and protection equipment. A couple of process data are serving for global process supervision. Actions for the optimization of the system are mostly initiated manually through the centralized control system on the base of predefined network control schedules and results from several energy management applications. With the liberalization of the energy markets, the requirements for control centers are increased. At first, the power systems have to be operated more efficiently due to the increasing economic targets. Therefore the operational limits have to be determined more precisely. Second, the networks have to be operated more flexible, because for the network planning it is harder to consider more and more volatile market activities. The basic data for the planning and operation are changing faster and are partly more uncertain. Simultaneously, the progress in information technology allows a more and more flexible data transmission and processing, which leads to an increasing supply with process data, which have to be used in a profitable way.

Conclusively, the control and protection systems are getting more complex with changing and increasing requirements. The described autonomous system structure enables the consideration of all these requirements because of its capabilities. Especially the action capability of the system without manual interactions is extended and prolonged during fast system variations and events. Actions will be enabled based on uncertain and fuzzy data. Therefore, the process operator can concentrate on higher value decision and optimization tasks.

1.4.1 Operational Architecture

The transposition of the existing hierarchical structure of control systems with system or network control level, substation control level and bay control level into the structure of autonomous systems is shown in Fig. 1.8. The general functional architecture is transposed into the operational architecture related to the control system of electrical power systems. The protection system is integrated within this structure as well. The particular levels are represented by several autonomous components for different tasks.

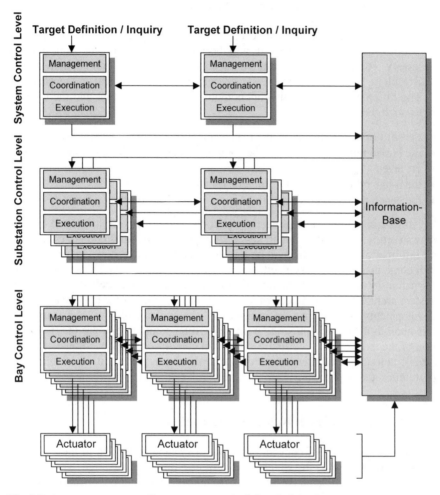

Fig. 1.8. Autonomous system for process control of electrical power systems

The actions and information of the components on the higher levels are transmitted via the information base to the underlying levels and defining or influencing their goal definitions. It has to be noticed, that the information flows are flexibly linked.

A component on the system level is communicating with the related components on the substation or bay level for which the goals are defined or for which the information is important. The autonomous system architecture provides in comparison to conventional systems both the flexible communication and complete modularization of the applications. This information flow and access is one novelty that autonomous systems provide, whereas nowadays lower hierarchical levels have no information access to the same or a higher level. The newly provided communication capability breaks up the hierarchy of information flows only from the bottom to the top.

A component on a particular level is able to access all information from all components in other levels or all neighboring components on the same level if necessary. The components act autonomously, if this information is not available. The resulting structure can be called as decentralized, hierarchical and meshed.

The differences between the autonomous system architecture and the conventional architecture are:

- the shift of the organization and management functions into lower levels,
- the horizontal coordination of the components on the lower and especially lowest level,
- the usage of all available information on all levels,
- the generation of high value information, which is also used on the lowest level.

All existing control and protection functions can directly be integrated into the new structure. Doing this, not all layers within the autonomous component and not all information channels have to be implemented. For future developments, which will come due to necessary intelligent decision techniques as well as complete information integration and communication interconnection, a well-structured open system architecture is defined, which enables an optimal implementation of these developments. Implementation examples will be explained in the later chapters of this book in detail. The integration of the existing solutions enables the transposition of the existing into the future system. In conclusion, the operational architecture of autonomous systems enables the structured implementation of future opportunities and applications.

1.4.2 Characteristics of Autonomous Systems

The bay control level of conventional control systems is limited to the execution of pre-defined actions, for which short response times are required. The network control level is dominated by functions for management and organization like e.g. planning and optimization tasks, which require human interactions and therefore have longer response times. The shift from organization and coordination tasks to lower levels enables the decentralization of several tasks, which reduces the spatial distance to the actors and therefore the communication effort. A fast event-oriented action scheme can be performed also in time critical situations. The response time for complex actions is getting shorter.

Actions are arranged in conventional systems on the highest level, which are considering longer time intervals and therefore intense variations of the process. The extension of the lower control levels with components for event orientation and the consideration of crosswise interconnections as well as high value information extend the time horizon for their automated actions. Manual interactions and modifications, which are defined on higher levels, are necessary only in longer time intervals. The very limited information context of today can drastically be extended by the complete availability of data, information of neighboring components, self-diagnosis and the generation of higher value information.

How to implement the structure of autonomous systems will be discussed in the following chapter, which leads to the implementation architecture.

References

[1]   Antsaklis PJ, Passino KM (eds) (1993) An Introduction to Intelligent and Autonomous Control. Kluwer Academic Publishers, Boston, Dordrecht, London

[2]   Lagerberg (ed) (1996) Organisation and Design of Autonomous Systems. University of Amsterdam, Faculty of Mathematics, Computer Science, Physics and Astronomy (WINS)

[3]   Passino K (1995) Intelligent Control for Autonomous Systems. IEEE Spectrum, June: 55-62

[4]   Abel E, e.a (1993) A multiagent approach to analyze disturbances in electrical networks. In: Proc. of 4th ESAP Conference, Melbourne, Australia, pp 606-611

[5]   Talukdar S (1993) Asynchronous Teams. Proc. of 4th ESAP Conference, Melbourne, Australia, pp 647-655

[6]   Talukdar S, Ramesh, VC (1993) Cooperative Methods for Security-Planning. In: Proc. of 4th ESAP Conference, Melbourne, Australia, pp 93-98

[7]   Rehtanz C, Becker C (1998) Autonomous Systems for Intelligent Coordinated Control of FACTS-Devices. Proc. of Bulk Power System Dynamics and Control IV - Restructuring, Santorini, Greece

[8]   Rehtanz C, Handschin E, Becker C (1998) Coordinated Decentralized Control of FACTS-Devices Applying Autonomous Systems. In: Proc. of International Conference of Electrical Engineering, ICEE '98, Kyongju, Korea

[9]   Fogel DB, Fukuda T, Guan L (1999) Scanning the Special Issue - Technology on Computational Intelligence. IEEE Proceedings Special Issue on Computational Intelligence, vol 87, no 9

[10]   Van Son PJM (1995) Practical aspects of "intelligent machines" in the control room. Electra 159, pp 102-116

2 Implementation of Autonomous Systems

Christian Rehtanz, Tatjana Kostic, Martin Naedele

ABB Switzerland Ltd - Corporate Research, Baden-Daettwil, Switzerland

This chapter introduces the Information Technology (IT) architecture of autonomous systems. Autonomous systems in this context realize the control system of electrical power systems. The basic implementation aspects of the architecture are described in comparison to conventional approaches. The nowadays' available methods and standards of IT are analyzed to show that the proposed approach can be implemented today. The topics addressed are communication systems, database systems, software services, interconnection protocols and international standards for data and application models as well as IT security issues.

2.1 Requirements for the Implementation Architecture

The previous chapter has introduced the functional architecture of autonomous systems, which defines the principal targets and capabilities of autonomous components and systems. The logical relation between autonomous systems and power systems control has been introduced as the operational architecture. An essential part of autonomous systems are the algorithms enabling the intelligent processing of information. These algorithms will be described separately in chapter 4. Another important aspect is the technical implementation called implementation architecture, which will be described in the following.

The basic requirement for the information technology and the implementation architecture of autonomous systems is the free data access and exchange between all autonomous components. The following section will not specify the implementation architecture in all details, but will clarify whether the basic requirements can be fulfilled with the information and communication technologies available today.

Another requirement is the usage of standard communication technologies, protocols and models, which will enable and promote the implementation of the autonomous systems architecture in practice. Several automation and IT standards exist. These standards are more and more used in power systems to enable cost-effective solutions for future control systems, and thus represent natural candidates for the implementation of autonomous systems.

The main drawbacks in today's control system architecture can be explained with the help of Fig. 2.1. The data and information exchange within the control systems of electrical power systems are structured into a strong hierarchy. Measurement values, or process data, are transmitted from lower to higher levels (bottom-up), while the control information is transmitted vice versa (top-down). The communication bandwidth within a station, and that between stations and the system control level are limited. Several vendor specific protocols are used with the need for specific gateways (GW). Fixed information channels, e.g. telecontrol systems, are supporting the hierarchy.

The applications within the Energy Management System (EMS) are not modularized and are based on the central SCADA database information. Interconnections to other information systems, such as enterprise information or maintenance management systems, have to be handled by specific interfaces. The SCADA system itself contains both real-time and configuration data. The Intelligent Electronic Devices (IED) like protection devices on the station and bay level are also vendor specific with varying functionalities, interface protocols and data models.

To avoid vendor specific and bandwidth limited communication, standard protocols and data models have to be applied. For example, dedicated communications channels are more and more replaced by communication means like field busses, local area networks (LAN) or wide area networks (WAN). Data models are defined in some upcoming standards.

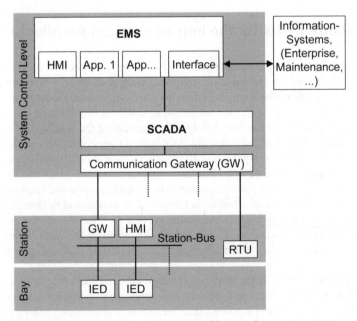

Fig. 2.1. Standard control system communication architecture in power systems

The IT architecture has to be provided for the three logical information channels that have been specified in section 2.2.4, namely, real-time data, high-value information, and diagnosis/function information. In contrast to the current hierarchical architecture, the new implementation architecture has to offer services that support the non-hierarchical system-wide information access, the real-time data exchange among components on the bay level of different substations, and an active sending (or pushing) functionality to other components. For all interactions of the autonomous components, real-time data interconnection as well as information or database access in all directions have to be provided. The increased local data processing, enabled by the operational architecture of the autonomous system, leads to an important decrease of the data traffic between the hierarchies, but increases the traffic within the levels. All these result in specific requirements for the implementation architecture.

The following sections discuss to what extent the standard technologies are sufficient to implement autonomous systems, and pinpoint the areas with needs for new developments. The topics discussed are communication protocols (2.2), standards defining data and application models (2.3), information systems (2.4) and IT security aspects (2.5) for the implementation of autonomous systems.

2.2 Communication Technology

From the perspective of the implementation architecture, the physical structure of an autonomous system is a distributed computer system [1]. The control system applications are component based software modules with distributed data storage within an open system and software architecture [2],[3]. The control system applications can be arranged according to logical and functional aspects thanks to the free and location independent data access.

The new communication architecture uses the Open Systems Interconnection (OSI) layer model for communications [4] as its reference model. This 7-layer model separates different communication tasks from the physical layer, over the data link, network, transport, session and presentation layers up to the application layer. Any communication protocol can then be defined in terms of profile, which specifies the layers the protocol supports.

The traditional communication architecture, described in the previous section, uses mainly three- or four-layers protocol profiles over dedicated links. These protocols were designed to support reliable communications on narrow-band serial data links traditionally used for communications between a SCADA master in a control center and RTUs (Remote Terminal Units) located in substations. The data exchanged, i.e., the measurements received or the devices controlled, are anonymous data points, translated into a number or address in the RTU.

The following description reviews standard protocols appropriate for the application to autonomous systems within the new communication architecture.

2.2.1 Topology

The basic topology of the implementation architecture is outlined in Fig. 2.2. Local area networks (LAN) are used for the data communication within a substation. The communication between substations is implemented with wide area networks (WAN) to which the substations are connected via gateways (GW). Some IEDs from substations without a substation automation system can be directly linked with the WAN.

Starting from the lowest layer of communication, the physical layer, the WAN consists of meshed point-to-point interconnections of copper cables, fibre-optic cables, radio links or satellite channels. The physical interconnections of the WAN are usually not exclusively available for the control system applications. Therefore special transfer rates or channels have to be reserved.

To fulfill the requirements concerning bandwidth reservation and security for real-time control systems, the WAN for the use in electrical power systems is separated from public WAN. Either completely separated communication links or separated channels of one physical interconnection establish the private WAN for the control system. Information systems like enterprise or maintenance systems, which are usually interconnected through office communication to the public Internet, have to be interconnected through information servers separating the private WAN from public areas.

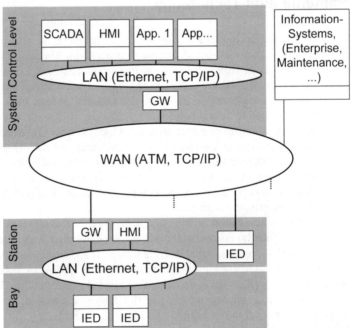

Fig. 2.2. Future communication architecture in power systems based on standard technologies

2.2.2 Protocols

The main requirement to the communication protocols is to guarantee well-defined latencies for the real-time data, while allowing an efficient transmission of non-real-time information.

The second layer of the communication model, the data link layer, can be implemented for example with a cell transmitting protocol like ATM (Asynchronous Transfer Mode), which is a compromise between connection-oriented and packet-oriented communications [5]. For instance, current broadband ISDN-systems in Europe are partly based on ATM. These existing standard systems can be used for the wide area communication of the implementation architecture.

Despite of its asynchronous multiplex operation, ATM can be used for synchronous communication because of a fixed length of cells. The information on the connection owning a cell, is included in the head of the cell. Hence, there is an explicit interconnection between source and sink, which can be called a virtual channel. This virtual channel has to be defined before the data transmission. All data cells are transmitted via this channel to the target sink, and the order of the cells thus remains unchanged. Communication partners define, according to their requirements, the Quality-of-Service (QoS) of the communication within ATM-systems through markers in the protocol. The QoS defined by the markers in the packages allows to distinguish, for example, between real-time measurement data and non-real-time maintenance information. The QoS is observed continuously by the communication system. ATM-systems achieve transfer rates up to 2,4 GBit/s.

The substations, which are interconnected to the WAN via routers or switches, can perform their local communication through a LAN. The Ethernet LAN protocol enables transfer rates up to 1 GBit/s. The switched Ethernet protocol is a quasi-standard for LAN. If sufficient bandwidth is provided within the switched Ethernet-LAN, the requirements for real-time communication can be fulfilled on this layer.

On the lowest level of process architecture is the field or process bus system. It is implemented on the bay control level to interconnect actuators, sensors and the field processor. Short data packages with measurement values, messages, alarms or control signals are exchanged within this communication system. The boundaries between field bus and LAN systems are more and more blurred, thanks to more powerful and intelligent actuators and sensors. The sensors are equipped with processors for the direct interconnection to different kinds of communication systems. A pre-processing of data can be implemented directly within these devices. A station wide LAN comprising the bay control level can be implemented today [6].

The next two communication layers, network (3) and transport layer (4), on the top of the ATM-WAN and the Ethernet-LAN can be implemented with the standard Internet technology TCP/IP (Transmission Control Protocol / Internet Protocol) [7]. TCP/IP links different kinds of communication networks to build logical point-to-point interconnections. It can be applied on the top of ATM and Ethernet as a universal protocol within the entire system. The next generation of the IP-implementation [8] will allow to define QoS also on this protocol level. This will enable not only to telephone via Internet, but also to set up real-time data exchange within the control system specified here.

Virtual channels within the communication system can be specified with desired QoS for the particular tasks. Therefore, the system-wide communication can be implemented according to the requirements. It must be noticed that establishing a virtual channel requires a certain time, and that it can fail if the communication system capacity is used to the limits. Therefore, the most important control system interconnections must be established during the initialization of the system, and then kept permanently. The supervision of these channels requires a frequent data exchange.

Beside these permanent channels, further connections between all processing units within the distributed control system can be established on demand. These connections enable the parameterization or status requests of particular components. For instance, if switching actions within the power system require a new communication channel, the channel has to be established before executing the switching. The autonomous components have the responsibility to establish these virtual channels.

Within this new architecture, the SCADA-system has no longer a central position. It will become an application or one autonomous component with the task to provide data consistency. Other applications can now access the consistent sets of data either from the SCADA-component or directly from the process. Therefore the hierarchy is broken up and the system is theoretically symmetric, as seen from its components. This leads to the free data-access provided to the autonomous components, as shown in Fig. 2.3.

The session layer (5) and presentation layer (6) of OSI model are usually not used if TCP/IP is the underlying protocol. These services enable for control system applications the access to data and information with simple commands. Examples for these services are the e-mail service (SMTP), the file transfer protocol (FTP) for the transmission of complete files or HTTP (HyperText Transport Protocol) in the Internet.

Finally, the application layer (7) of OSI model accesses the communication functionalities through high-level protocols providing services for particular communication tasks. MMS (Manufacturing Message Specification [9]) is such an application layer standard. It defines data types and a common message format for providing a wide range of services to the real-time applications. Some of the services include, reading, writing, and reporting of variables (simple or arbitrarily structured data types), event management, journaling, remote program control, and uploading/downloading of data and programs.

The MMS protocol was originally developed for the production within the automotive industry. Since it provides a rich real-time network-programming environment to support a very wide range of distributed applications, it has already been adopted as the basis for service definitions in other automation areas, including utility control systems.

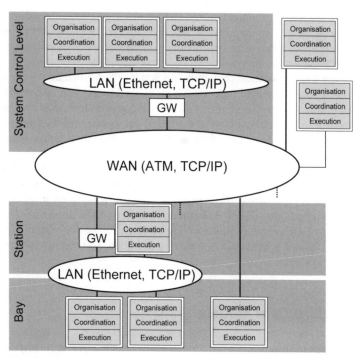

Fig. 2.3. Concept of the communication architecture of an autonomous system

There are notably two utility-specific communication standards which use MMS as application layer protocol. The first one is specified in Part 6 of IEC 61870: Telecontrol equipment and systems [10], also known as TASE.2 or ICCP (Telecontrol Application Service Element 2 and Inter-Control Center Protocol, respectively). Its purpose is to provide the means for data exchange between control centers. In addition to SCADA data and device control, TASE.2 also provides for exchange of information messages (i.e., unstructured ASCII text or short binary files) and structured data objects, such as transmission schedules, transfer accounts, and periodic generation reports. The second example of standard using the MMS protocol is IEC 61850: Communications networks and systems in substations [11], which is currently being developed. This standard draft defines abstract communication service interfaces (ACSI) and then provides their mapping to the MMS protocol (among others). The IEC 61850 will be described more in detail in the next section, in the light of its data model.

Besides the established control system protocols, it is possible to simultaneously use the modern text-based and web-service-oriented process control stack OPC-XML / SOAP / HTTP / TCP (OLE for Process Control [12] - eXtensible Markup Language [13] / Simple Object Access Protocol [14] / HyperText Transport Protocol / Transmission Control Protocol [15]). These protocols allow a transparent set up of services and applications, but include a very high protocol overhead (XML is inherently quite verbose), which means they have high requirements for the communication bandwidth. All applications based on non real-time data can be implemented using this standardized way. With the increase of the available bandwidth, it is conceivable to apply these protocols also for real-time applications in the future.

All of the communication protocols and services described above are standards with either a broad installed base already now, or represent an upcoming technology in the near future. Their use will lead to the solutions that are more cost-effective than those deploying vendor specific protocols, and will enable flexible and modularized systems. The obvious pre-requisite is that device manufactures and system vendors implement these industrial standards, which requires a certain effort. Implementing standards basically translates into developing software applications for both devices and systems, that will let those devices and systems support the services specified in the standards. The following section will discuss the emerging standards that define object models of the utility domain, which greatly facilitate the development of software applications for power system control, and provide for intuitive semantics of data exchanged among heterogeneous and autonomous systems.

2.3 Standard Data Models

The communication protocols introduced in the previous section "physically" enable data and information exchange among autonomous systems in real-time operation. This section will focus on the data models specified in the standards, which provide for more abstract means for enabling communication (i.e., exchange of data), for both real-time operation and non-real-time tasks such as system configuration or inter-application communication (i.e., message sending).

Traditional and still widely used protocols usually implement only the lower layers from the OSI reference model (see Section 2.2). The data model in this case is simply an array containing signal addresses (or data points), and there is no notion of devices at the protocol level. This type of communication is an anonymous exchange of data points, because a physical device that receives the value knows neither what the meaning of the value is nor which physical device sent it - it only knows that the signal comes from a given address. The same comments apply to the sending device. To give semantics to the values used by the applications at any end of the communication channel, there must be a mapping facility at each end that has to be configured so as to associate data points with meaningful physical objects (e.g., circuit breaker) and their attributes' values (e.g., status = "open").

The following sections will briefly introduce three emerging IEC standards that define more elaborate data models. Their application can eliminate the shortcomings of point-oriented, anonymous data exchange, and enable direct interaction among autonomous components. Although all these standards are still only in draft versions, some of their parts have already been adopted as *de facto* standards and even implemented by some device manufacturers and system vendors.

2.3.1 Substation

The IEC 61850 standard [11], introduced in section 2.2.2, makes a big step forward with respect to data model for substation systems and devices. Although it is basically a *communication* standard, great effort was invested in the domain analysis. This resulted in quite an elaborated domain model, which contains also the data model. The main abstraction of the *domain* model is the Logical Node (LN), which can be seen as an atomic functionality available within the substation, from the substation control system, to protection and control devices, to the process itself. An LN holds data, classified into a number of Common Data Classes, which are the main abstraction of the *data* model. Each physical device contains a number of LNs, and may act as a server to any other physical device or system. Functionality of devices and their interactions are realized through collaborations among their LNs.

Logical Nodes and their data are, unfortunately, specified as tables in Word documents of the standard, although there are more formal methodologies to unambiguously specify such models. Nonetheless, there are efforts to provide a formal model of IEC 61850 [16].

As stated above, each LN encapsulates the data it needs for performing its functionality or behavior. (Note that the behavior is not defined in terms of operations, but is implicitly contained in the definition of LNs and the values of their data attributes). An important fact is that the data contained in LNs is not only typical operational data (e.g., measured value or position status with their quality and time tags), but also different configuration data (such as the device nameplate or its last configuration version). This implies that the devices can describe themselves.

In the operations context, devices are servers and clients, and thus can be queried or query other devices through ACSI services. For configuration purposes, the IEC 61850 defines Substation Configuration Language (SCL). SCL is an XML Schema, with the elements and attributes reflecting the domain model. So, the self-description capability of IEDs can be made available in a standard, human- and machine-readable way, through an XML instance file. Additionally, SCL allows one to configure the communication-related attributes of an IED (e.g., supported ACSI services, server address), as well as to describe the equipment and communication topology within the substation.

The concept of abstract LNs, which model atomic behaviors within devices and systems, encapsulate own data, and collaborate with other LNs through server-client mechanism, can facilitate implementations of the IEC 61850 with the means of software components, as proposed in [17]. This kind of approach is an illustration of autonomous components.

2.3.2 Network Control Center

Another IEC standard in progress, IEC 61970: Energy Management System Application Programming Interfaces (EMS-API) [18], defines the means that can facilitate "opening" of traditionally closed and monolithic EMS/SCADA systems, which equip the network control centers. The standard defines the *data* model, called CIM (Common Information Model), and a set of application programming interfaces (APIs) used to manipulate the EMS/SCADA database data.

Contrary to the IEC 61850 for substations, this standard does not specify any particular communication protocol. It rather specifies concrete APIs (operation signatures in different programming languages), which then can be realized by a protocol with the OSI profile appropriate for the given execution environment. There are two global sets of APIs. One set is intended for fast access to real-time data, typically used within SCADA applications, such as a data acquisition front-end or an update of human machine interfaces with the process data. The other set of APIs enables near-real-time access to the full network model or its parts, typically used in EMS on-line applications, or even out of the EMS for, e.g., long-term planning or reliability analysis applications.

A notable contribution of this standard is its data model. CIM is an abstract model that describes the domain known to EMS/SCADA systems as a set of objects with attributes and relationships to other objects. The model is defined in UML (Unified Modeling Language) [19], a *de facto* standard not only in computer engineering community, but also widely used for general domain analysis or modeling of business processes. The CIM model is maintained with the help of a CASE (Computer Aided Software Engineering) tool, which also allows to automatically generate the model documentation (i.e., the standard itself!) and keep it consistent with the model.

Similar to the IEC 61850 for substations, this standard also defines the serialization format, i.e., the format for instance files that contain the network description and configuration data. The CIM/XML language [20] is an application of RDF (Resource Description Framework [21]). Its syntax is XML, and its semantics is based on CIM objects and their attributes and relationships. The RDF schema can be automatically generated from the CIM UML model with an open source software tool. The CIM/XML has been set by NERC (North American Electricity Reliability Council) as the format for network model exchange among the North American transmission network and system operators. CIM contains thus the so called "NERC profile", which tags the elements of the CIM UML model that must be present in instance files for network model exchange.

CIM and the APIs defined in IEC 61970 represent obvious means that support modularizing of EMS/SCADA systems, as required by the autonomous systems architecture.

2.3.3 Electrical Utility Enterprise

The last standard that will be mentioned here, IEC 61968: System Interfaces for Distribution Management [22], is also the work in progress, whose development is closely related and well coordinated with that of the IEC 61970 (see Section 2.3.2). It is designed for Distribution Management Systems (DMS), which typically must communicate with network operation and process control systems

(EMS/SCADA and substations, respectively), as well as with different enterprise level systems, such as customer management, resource planning or maintenance and outage management systems. Therefore, the objective of this standard is to define a set of messages that DMS needs to exchange with other systems within the enterprise.

This standard is complementary to the IEC 61970 (EMS/SCADA) for two reasons. First, it uses CIM as its domain model and extends it with the physical objects and concepts that are relevant for distribution networks only. Therefore, all the above discussion on CIM applies here as well. The second reason is that this standard also uses the APIs defined in IEC 61970, where applicable, and extends them with APIs that allow inter-application messaging. This means that these extended interfaces can be used by EMS/SCADA systems, as well.

Messages in IEC 61968 are specified in XML. Since the time scale of inter-application messaging is not real-time critical, the implementations are likely to simply use some XML-based protocol, such as SOAP, or execution environments and programming languages, which provide native support for XML, such as CORBA or Java.

2.3.4 Summary

The emerging standards, which define data models that carry domain semantics, can find their application to different stages within the lifetime of an autonomous component and the system where the component lives (design, commissioning, configuration, operation). Some of the standardized data models, in particular those that are *object models*, offer several obvious advantages over traditional data point representation of data. They inherently contain the domain semantics, which are widely accepted and agreed upon among the domain experts. The devices and the systems offered by different vendors and manufacturers can all be described and configured with the same semantics, thus becoming pluggable into any foreign system that understands the data model. Thanks to the common semantics, application engineers can comfortably develop applications, without worrying about the protocol details or, e.g., the unit or the source of the value - these are implicitly contained in the model. With the models expressed in standard modeling languages (such as UML) within the CASE tools, the tasks of software engineers that are responsible for the implementation of the standards in devices and systems are facilitated and less error-prone. For instance, the skeleton source code or the documentation can be automatically generated with the CASE tool. Finally, the adoption of standard data model opens the way to develop new applications that extend existing systems, thus providing more choices for the utility while securing their investments into the current systems.

Although all the three IEC standards described above are still in draft version, some of their parts have already been implemented by system vendors and device manufacturers. All together, they provide the utility control system domain model, from the process bus in substations up to the enterprise-wide applications, and thus can contribute to the implementation of autonomous components that can communicate seamlessly.

2.4 Information-Base of Autonomous Systems

After the specification of different aspects of the communication system and data and application models, the data administration will be discussed in the following. Large amounts of data at different locations are processed and provided by geographically distributed autonomous components. The basic requirement for the data and information storage is to avoid inconsistent or even redundant data sets throughout the system. The structure of autonomous components shall be used as far as possible to keep the flexible and modularized structure.

The storage of data and information in the autonomous system is logically a distributed database. In comparison to database servers, which are administrating the local data and building a functional data separation, distributed database systems administrate the data physically stored at several points, system-wide. The development of distributed databases is going on rapidly simultaneous with the trend to distributed computer systems [26]. Such a database system is overlaid to the proposed communication system and stores on every level the accrued data like operations logs, load profiles, energy management data, maintenance data, etc. For applications, there is no difference between centralized and decentralized database systems. An application within an autonomous component is requesting data or information from the system and will get the required service from another autonomous component. The autonomous component will either provide the information from an internal storage or through an online calculation.

One specific aspect of this data and information provision is the integration of different information systems even outside the power system control like enterprise information systems for maintenance scheduling, energy market, customer handling etc. Enterprise information systems as such are growing together through the integration of different databases and information sources. Technical and economical data will be integrated into common systems to administrate and use information as a resource. These information systems allow the access to all relevant information. They are supporting in this way the decision process within an enterprise. Especially the enterprise data in this context include also the energy market data, which are more and more decisive for the technical process and vice versa.

The usage of such an information system can be extended in a way, that the process or the process operator is able to get not only its own, but also all other related and relevant information. This information access must be supported by the information access of the autonomous component. All enterprise information systems seen from the autonomous component point of view are other autonomous components. Their information content can be received and sent on demand like any technical information according to specifications of content and time-ranges. This approach is similar to the Web-service philosophy, which also enables the access to information for which someone has access rights or is subscribed to. All in all, this information interface seen from the autonomous component enables the direct access to technical process data, like the power system control data and information, as well as all kinds of enterprise data. Therefore, at all points of the process all data and information are accessible in time.

2.5 IT Security Considerations

When discussing and designing a communication, data exchange and application architecture for the systems, which control or are being used to control elements of a country's critical infrastructure, IT security aspects and concerns have to be taken into account. The threat and danger of IT network-based attacks on critical infrastructure installations by terrorist or criminals have become clear to everybody, not least due to recent geopolitical developments.

In the previous sections it has been stated that autonomous control systems for power systems will communicate via WAN and LAN networks. Even if this is not the public Internet, but a company internal, 'private' WAN, it has to be regarded as public from the security point of view, as there is, by definition, a multitude of different users, and applications interacting with each other over which no single party has complete control. Any potential security benefits of such a limited user community disappear, if the control system information has to be made available to enterprise systems, which are usually interconnected to public networks for office communication tasks.

IT security is more than the issue of confidentiality alone - and thus encryption is not the cure-all. It also encompasses the aspects integrity, auditability, non-repudiability, access control, and availability of information and services.

While from a functional and economic point of view very beneficial, the current trends towards total interconnectivity and open, standards-based protocols, applications and devices also cause a higher risk of vulnerability to network based attacks:

- Attacks can be conveniently executed from any part of the world.
- Legal prosecution is not a useful deterrent, e.g. because the attacker cannot be identified, or his actions are not a punishable crime in the country where he is based.
- Technical details and vulnerabilities of these open systems are known to more people than of their proprietary predecessors.
- Interoperability and interoperation between applications and devices of different vendors within the autonomous system prepares the ground for deliberately and incidentally malicious interactions.

These facts require that IT security concerns are taken into consideration already early in the design process for an autonomous system. While a thorough security needs and means analysis, design, implementation and operation can not guarantee one hundred percent security - in the same way that no amount of safety measures can guarantee perfect absence of any accident - already with today's means IT systems can be built that offer reasonable risk/cost/protection trade-offs.

Left to themselves, all IT systems evolve over time towards weaker IT security. This is not just true of the procedures involving humans, but also the security applications themselves offer less and less relative protection, e.g. due to discovery of new vulnerabilities in deployed applications, development of new attack types, and more computational power available to the attackers for less money (Moore's Law). Nevertheless, the concepts of *time-based security* and *defence-in-depth* allow us to achieve a high level of security despite the fact of imperfect security system elements.

Time-based security [23] can be summarized in equation (2.1), which states that the protection in terms of time an attacker is delayed in getting to the valuable parts of the system (*P*) must be at least equal to the sum of the time needed to detect the attack (*D*) and to react effectively to it (*R*).

$$P \overset{!}{\geq} D + R \tag{2.1}$$

Of course, neither the left nor the right part of this formula are constant and clearly determinable, but require good human judgment during analysis and design. The positive content of this statement is that P does not have to be infinity, if D and R are sufficiently low. And this leads to the second concept mentioned above: defense-in-depth.

Traditionally, there are two main approaches to designing 'systems with security': The first one, named 'hard perimeter', takes the functional system as it is without security considerations and then wraps it with a more or less impenetrable security perimeter. The advantage of this approach is its conceptual and implementational simplicity. The disadvantages are exemplified by its most famous historical example case, the Chinese Wall:

- Without detection mechanisms outside the protected zone, an infinitely strong defense is required.
- Experience shows that, for human and technological reasons, it is impossible to design and implement an infinitely strong defense.
- Especially the gates necessary to deliver the desired functionality are in the operational phase vulnerable to social engineering attacks.
- Monoculture is dangerous: The defense is based on one principle or platform; if this fails for some reason, the whole defensive capability is gone.
- Once the attacker has broken through the defense at one point, no more security reserves are left and he owns the whole system.
- No protection is provided against malicious insiders or authorized applications with malicious side-effects (Trojans).

The other approach, 'defense-in-depth', uses a variety of defense, detection, and reaction mechanisms to divide the system to be secured into compartments which can be defended individually and which are arranged such that an attacker has to successfully subvert less valuable and critical compartments first before getting to the really sensitive systems. This delay allows initiating a successful response. The historical parallel for this approach is the medieval castle. In this approach the security mechanisms in and around each compartment are a mix of dedicated devices and applications as well as mechanisms, such as role-based access control and application level anomaly detection, directly built into the applications providing system functionality. Thus a system following this approach is, of course, more complicated and, superficially regarded, costly to understand, realize, and maintain, especially if security has not been a consideration from the beginning of the design of the system. Nevertheless, only this approach can, if appropriately implemented, deliver a really secure system.

The discussion so far leaves open the question of how to build the communication and application architecture for an autonomous control system which implements the defense-in-depth approach and uses today's security mechanisms.

There is not enough room here for a more detailed treatment of the quickly evolving technical details of the various protection mechanisms. Various books on designing secure systems and their elements exist, e.g. [24]. Reference [25] surveys today's IT security mechanisms on the basis of a generic security zone model - both generic ones and those specifically applicable for automation systems, because they make use of the particular characteristics of automation systems. Such a defense-in-depth security system will for example contain:

- Multiple firewalls in series and for compartmentalization of the communication network, diverse in functional principle (filtering routers and application level gateways) and vendor/platform.
- Hardened hosts as well as services and data flows restricted to those essential for delivering functionality.
- Encrypted and mutually authenticated communication.
- Application proxies to isolate server applications with known vulnerabilities or complex behavior.
- Role-based access control designed into functional applications.
- Intrusion detection systems on each firewall-limited segment of the network with centralized automated log consolidation and analysis application.
- Operation centers staffed around the clock with IT security experts to respond to alerts with appropriate reactive measures.

A simplified generic topology for a network comprising both enterprise and office applications is shown in Fig. 2.4. It should be mentioned that the private WAN and the public WAN resp. Internet could physically be the same communication lines, but using different, strictly separated channels.

Critical access paths between enterprise applications or also non-time-critical information exchange between different control centers can be implemented as virtual private network paths supported by standard encryption methods.

As a conclusion it can be said that the security aspects of confidentiality, integrity, auditability and access control are very well manageable with today's mechanisms deployed according to a defense-in-depth strategy. The main open problem for wide-area, real-time or near real-time communication systems such as those required for autonomous power system control is availability and thus the protec-

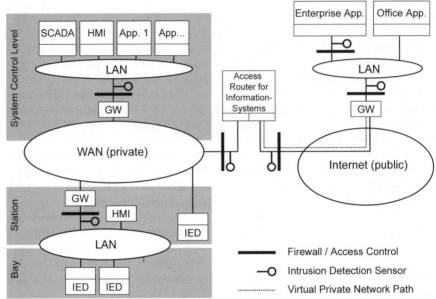

Fig. 2.4. Topology of autonomous control system architecture regarding IT security

tion against denial-of-service attacks. This requires a careful design involving, among other things, priorities, filters, redundant and backup communication systems and paths as well as the capability of parts of the control system to operate in isolation.

2.6 Conclusions

In conclusion, the entire concept of autonomous systems can be implemented using existing information technology for communication and data modeling as well as for data and information access, all in a secure manner.

The open question of how to use all these new data and information in a profitable way and how to implement decentralized procedures will be clarified in the following. The next chapter introduces some basics about knowledge processing and computational intelligence together with methods, which are able to generate high value information out of available basic data. Such methods of computational intelligence are building the background for the implementation of autonomous systems, whose application examples will be discussed in detail in the application chapters of this book.

References

[1]   Fleischmann A (1994) Distributed Systems - Software Design & Implementation. Springer

[2]   Handschin E, Nikodem T, Palma R, Rehtanz C (1997) Object-Oriented Based Methods and Techniques in Control Centres Applied to Open Access Schemes. Proc. of 10th International Conference on Power System Automation and Control, PSAC '97, Bled, Slovenia

[3]   Bose A, Green TA (1992) Open Systems Benefit Energy Control Centers. IEEE Computer Applications in Power, vol 5, issue 4, pp 45-50

[4]   International Organization for Standardization (1984) Basic Reference Model for Open Systems Interconnection, ISO 8072

[5]   Miller A (1994) From here to ATM, IEEE Spectrum, June, pp 20-24

[6]   Banhan D, Hill K (2000) The fusion of Internet and Ethernet Technologies ushers in a new era in Substation Automation. Conference Volume "Monitoring and Control by Internet", ERA Report 2000-0102, ISBN 0.7008.0713.6

[7]   Stevens WR (1994) The Protocols - TCP/IP Illustrated. vol 1, Addison-Wesley

[8]   Loshin, P (1999) IP v6 - Clearly Explained. Morgan Kaufmann Publishers Inc.

[9]   ISO/IEC 9506: 1990, Manufacturing Message Specification (MMS)

[10]   IEC 61870-6: Telecontrol equipment and systems - Part 6: Telecontrol protocols compatible with ISO standards and ITU-T recommendations, 2001/2002

[11]   Draft IEC 61850: Communications networks and systems in substations, Nov 2002

[12]   Iwanitz F, Lange J (2002) OPC - Fundamentals, Implementation and Application. 2nd Edition, Hüthig Fachverlag, Germany

[13]   Gudgin M, et al (2002) SOAP Version 1.2 Specification. Parts 1 and 2, XML Protocol Working Group, W3C

[14]   Tidwell D, Snell J, Kulchenko P (2001) Programming Web Services with SOAP. O'Reilly & Associates Inc.

[15]   Perez J, et al (2000) Browsing through a distribution network. Conference Volume "Monitoring and Control by Internet", ERA Report 2000-0102, ISBN 0.7008.0713.6

[16]   T. Kostic, O. Preiss, Ch. Frei, "Towards the Formal Integration of Two Upcoming Standards: IEC 61970 and IEC 61850", to appear in Proc. of LESCOPE 2003 Conference, Montreal, 7-9 May, 2003

[17]   Otto Preiss, Alain Wegmann, "Towards a Composition Model Problem Based on IEC61850," Journal of Systems and Software, Special Issue on Component-Based Software Engineering, Elsevier Science, 2003, to appear

[18]   Draft IEC 61970: Energy Management System Application Programming Interface (EMS-API), Oct. 2002. Available: ftp://epriapi.kemaconsulting.com (documents), http://www.wg14.com (model view)

[19]   OMG Unified Modelling Language Specification, Version 1.4, Sep. 2001. [Online]. Available: http://www.omg.org/uml

[20]   Arnold deVos, S.E. Widergren, J. Zhu, "XML for CIM Model Exchange", Proc. IEEE Conference on Power Industry Computer Systems, Sydney, Australia, 2001. [Online]. Available: http://www.langdale.com.au/PICA

[21]   W3C, RDF Primer, http://www.w3.org/TR/rdf-primer/, W3C Working Draft, April 2002

[22]   Draft IEC 61968: System Interfaces for Distribution Management, May 2001. Available: http://www.wg14.com (no document download, only model view)

[23]   Schwartau W (1999) Time based Security. Interpact Press.

[24]   Northcutt S, et al (2002) Inside Network Perimeter Security. New Riders

[25] Naedele M (2003) IT Security for Automation Systems - Motivations and Mecha-
      nisms. atp, vol 45, no 5
[26] Öszu MT, Valduriez P (1999) Principles of Distributed Database Systems. 2nd Edi-
      tion, Prentice Hall, New Jersey

3 Computational Intelligence and Agent Technologies for Autonomous Systems

Christian Rehtanz

ABB Switzerland Ltd - Corporate Research, Baden-Daettwil, Switzerland

This chapter introduces the areas of computational intelligence and agent technologies for the application within autonomous systems. The key topic is the modeling of knowledge to establish intelligent behavior of an autonomous component, which also can be described and implemented as an agent.

An overview about different kinds of knowledge is given. Suitable methods of computational intelligence are explained concerning their special forms of knowledge. Most well known and some advanced methods are introduced. For theoretical details it is referred to the literature. The main characteristics of agent technologies are explained in detail considering their use in autonomous systems for power system control and operation. Specific characteristics of agents will be introduced where needed in the application chapters.

This chapter focuses on the selection of a suitable method if a particular kind of knowledge is available, which shall be integrated into the autonomous system structure.

3.1 Intelligence in Autonomous Systems and Agents

Methods for storing and processing knowledge in a computerized form are necessary for the implementation of the autonomous system structure regarding independent and intelligent behavior. Essentially, action schemes, which are thought up by humans, shall be mapped on a computer system, so that the behavior of the system can be called intelligent.

The methods of computational intelligence serve to make knowledge of different kinds usable for process control, operation and supervision. The question, which has to be clarified in this context, shall be less the methods itself but their capabilities and limits, which are leading to their selection for different applications. The application of the methods to solve technical problems has priority over theoretical topics of development and optimization of these methods, which are dealt with in special literature references related to the methods in this chapter. The following part targets on the principle aspects of mapping knowledge into technical systems and helps to select a particular method by knowing their validity and limitation for a particular problem.

The entirety of these methods is labeled with a couple of more or less synonymous names. Beside 'Computational Intelligence' also 'Artificial Intelligence', 'Knowledge Based Systems', ' Soft Computing' or 'Expert Systems' are used. All these acronyms have in common the integration of knowledge into a technical system or process. The difference between these methods is their kind of stored and processed knowledge. Generally, these methods are generating higher value output information out of input data with the help of stored and integrated knowledge. Which kinds of knowledge can be distinguished and which cognitive models can be derived from that, will be explained in the following section. After that the different methods are evaluated according to their kinds of knowledge.

3.2 Kinds of Knowledge and Cognitive Modeling

The different phenomena of knowledge can be described as kinds of logical correlations. It can be distinguished according to [1] in:

- certain knowledge,
- fuzzy knowledge,
- probabilities,
- plausible knowledge and
- experience.

The difference between these kinds of knowledge can be explained with an example of the maintenance of an electrical high-voltage breaker. The breaker has to disrupt a flowing current by establishing an isolating distance resp. the separation of two contacts to a certain distance within a certain time, while the arising electric arc has to be quenched. Breaker faults can occur for instance due to worn out contacts, fraction of high voltage parts or gas leakages of the isolated enclosure.

The fraction of a part of the breaker leads to an immediate maintenance demand. This knowledge is unambiguous and certain. As well, the mathematical function of a limiting value for the time to establish the isolating distance is certain knowledge. All the other kinds of knowledge, which are explained in the following, belong to the group of uncertain knowledge. In these cases the kind of knowledge is certain, but the knowledge itself contains uncertainty to a distinct extend.

The above-mentioned mathematical function for the isolating distance can be formulated considering the uncertainty like the following, if the limiting value for the time is not known exactly: "Maintenance is needed if the time for establishing the isolating distance is too long". The knowledge resp. the logical correlation is certain, but the formulation "too long" is fuzzy. In this case, the information about this correlation is taken from the process or a process model.

Another kind of knowledge is available in the form of data. Datasets can be evaluated, if they are available about the relation of process parameters, like for example for the establishment of the isolation distance related to the disturbances of the operation and therefore the maintenance demand. The statistical relations between the datasets can be determined with probabilistic methods. But the results are not leading to concrete recommendations for actions in the sense of a maintenance measure to be executed for a particular device. The hypothesis for a maintenance demand of a particular device is getting more plausible, if further hints for a maintenance demand are taken into account. Considering the uncertainty and plausibility of the knowledge, indications, supporting or contradicting the hypothesis of maintenance demand, are combined.

In practical applications there is usually a huge number of input information connected to desired output information via a complex relation. Despite of the complexity, this relation is known by experience. Experienced human operators can map a particular combination of measurement values and impressions, which means a specific data pattern, to concrete statements. This pattern recognition detects and evaluates the structural characteristics of the data.

The translation of all these kinds of knowledge phenomena into cognitive models of information processing or intelligent systems is targeting on the generation of concrete output information or actions related to particular inputs. If such a model is used within an autonomous component, an appropriate action has to be taken or proposed under certain conditions and with a specific target.

We can compare this technical knowledge modeling with knowledge representing models of human behavior. The cognitive model of acting humans corresponds to the description of their behavior through specific subjective reasoning and decision processes. This is the same behavior, which we want to implement in the autonomous component. In comparison to that, the average behavior is modeled with a sociological-statistical model, which can only be one hint in the direction of a specific decision or action. The methods, which are evaluated in the following, model different kinds of knowledge and are helping to represent intelligent behavior by their implementation into a technical system.

3.3 Methods of Computational Intelligence

3.3.1 Basic Methods

The basic methods for the implementation of intelligent systems are 'data based supervised methods', 'self-organized methods' and 'process knowledge based methods'. Additionally special optimization methods belong to this category. Basically such methods specify and parameterize a process or non-linear function F with parameters p, which maps inputs x to outputs y and represents the knowledge as shown in Fig. 3.1.

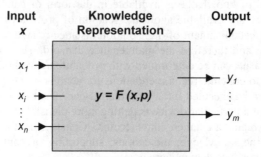

Fig. 3.1. Input-output relation represented by a knowledge based method

In this context the terms self-organization, adaptation and self-learning have to be distinguished. Self-organization is related to the processing of input knowledge, e.g. sets of input vectors x, without defining a target for desired output information y. Adaptation means, that a method resp. function F or its parameters p are adapted e.g. to data. Generally, a learning procedure adapts the information storage F according to the knowledge to be learned. In addition, the term 'self-learning' is used if this learning is performed frequently or permanently and not only ones to adapt the knowledge continuously to varying situations. All of these methods have already been assessed for power system applications [2]-[6].

The first group of methods are the data based supervised ones. The knowledge, which is brought into the process, is coming from the input data x with related desired or known output information y. The relation between input and output data represents usually the entire or a part of a technical process, which is unknown in an explicit form and is derived implicitly from the available datasets. The quality of the representation is evaluated resp. supervised during this procedure with the help of the data sets (x,y). The basic characteristic of these methods is that they interpolate output values related to similar but so far unknown input variables. Neural network like multi layer perceptrons ([7],[8]) belong to this group as well as decision trees [13]. Both of them are well known and power system applications are described in detail in the literature [7]-[12],[13].

The second group are data based self-organized methods, which are processing and structuring sets of example input data x. This unsupervised learning determines clusters resp. structural characteristics within the input data space. The characteristics of such clusters can be analyzed in a second step and be stored related to them. In operation, this method maps incoming data into a particular cluster or within the determined structure. The cluster or structure information can be used to evaluate the incoming data. The most well known method within this area is the self-organizing map [14],[15].

In the methods before, the knowledge was used from datasets. An unknown, complex process representation is derived from only input or input and output data. The application of these methods takes place if information about inner process relations is missing. Vice versa, the evaluation of the process itself is very restricted or even impossible with the help of these methods.

In the case of process knowledge based methods, information about the process is mapped into a specific representation. The influence of single input values on specific outputs can be formulated with qualitative or quantitative statements. As well if information about parts of the process sequence and their effects are known, statements can be formulated from which desired outputs or information can be derived. In these cases a couple of single input-output relations or relations between inner variables of the process serve to model a complex input-output behavior also called *F*. The input-output-representation can be used for a controller, as process model or information component. The knowledge is represented with statements in the form of rules, which are formulated either fuzzy or definite. If the input has a certain characteristic it can be reasoned that the output has a certain behavior for example in the sense of a statement to be taken or an action to be proposed.

The most well known representatives of these kinds of methods are fuzzy and expert systems. In fuzzy systems sets of parallel fuzzy rules represent direct input-output relations. Expert systems consist of chains of rules for the input-output relation. The rules are formulated definite in the simplest case. Several applications for power systems are introduced for fuzzy systems in [16]-[21] and expert systems in [22]-[24].

A method of indirect knowledge representation is the optimization. An optimization procedure serves for transposing an input request to an optimized output. The input request has to be formulated or integrated into the target function. The choice of the optimization method depends on the representation of the process, which has to be optimized.

If the process can be represented by a closed mathematical target function, standard optimization methods can be used. If the knowledge about the process is represented with alternative methods, e.g. methods of computational intelligence, special optimization methods have to be applied. Especially the quality measure for the optimization has to be handled particularly if it is formulated in a fuzzy or uncertain way. This is important for technical processes where a closed mathematical quality measure cannot be formulated, but the knowledge for the performance assessment is available. The quality measure can be formulated in this case with a fuzzy system, which can be handled by special optimization methods.

Optimization methods for such cases are Monte-Carlo-Methods, genetic algorithms and evolutionary strategies [25],[26], which are independent from the kind of quality function. In these methods several numbers of valid solutions are generated. The selection of the methods depends on the computation time for their generation. Power system applications are proposed and analyzed in [8] and [27].

Table 3.1 summarizes the basic methods of computational intelligence and evaluates them concerning their advantages and disadvantages for practical applications.

Table 3.1. Characteristics of basic methods

Kind of Knowledge	Method	Advantages	Disadvantages
data based, supervised (input and output datasets)	Perceptron Networks	- robust against input uncertainties - fast after training	- topology hard to define - high training effort - process knowledge not extractable from the trained network
	Decision Trees	- fast after training - transparent knowledge representation	- always a number of wrong classifications - limited robustness against uncertainties
data based, self-organized (input datasets)	Self-Organizing Maps (SOM)	- robustness of clustering - mapping of high dimensional data space to low dimensional output map - visualization capability	- no deterministic training results - strongly dependent on dimensionality of problem
process knowledge based (process relations)	Fuzzy Systems	- user friendly, transparent knowledge representation - robustness	- no security against implementation of wrong process knowledge - validation and tuning after basic design
	Expert Systems	- user friendly, transparent knowledge representation - ability to model highly complex relations	- high effort for complete and reliable knowledge representation - limited robustness
optimization methods (target function, quality function)	Genetic Algorithms / Evolutionary Strategies	- flexible formulation of quality functions	- no analytically proven optimum - problem must be formulated in a way avoiding invalid solutions

3.3.2 Extended and Alternative Methods

The methods introduced so far are theoretically deeply analyzed. Applications of these methods are known for several areas. Some drawbacks of these basic methods are limiting their applicability. Because of these restrictions, combinations of basic methods lead to applicable results in technical processes. In these cases all kinds of knowledge, which are necessary for the single methods, have to be available for the combined methods.

Beside the simple combination of methods, of which the application depends on specific problems, hybrid methods have been developed. These are neuro-fuzzy-systems and evolutionary neural networks [28].

Furthermore single methods have been improved to achieve characteristics like e.g. self-learning or flexible adaptation to varying data spaces. Self-learning and adaptive learning methods are adapting themselves permanently to incoming data. This adaptation goes beyond the simple self-organization of the learning process itself.

The basic methods presented in the section above differentiate strongly between learning and operating phase. In technical applications these methods can only be adapted through learning if datasets within an operational interval are stored and used separately in a learning phase. The newly trained parameters are handed over to operation frequently.

If self- or adaptive learning is necessary for an application, this has to be embedded directly within the method itself. Therefore, self-learning self-organized maps are introduced [29], which follow permanently varying structural characteristics or clusters of the input space. Other methods are neural gas and growing neural gas [30],[31], which also adapt to the structure of the input space. Both methods are data based, self learning and self-organized.

For the application of self-learning methods it must be defined to which extent already learned knowledge is overwritten by the adaptation to new knowledge. The importance of past information in comparison to actual inputs is a particular kind of knowledge, which has to be taken into account. The entire dataset from the past period can be stored in characteristic values, which are modified by new incoming values. This reduces the processing effort in comparison to methods dependent on extensive learning datasets.

A last alternative method is the evidence theory [32],[33]. This theory enables the combination of different kinds of information about a process to an output statement. The evidence theory is a more general description of the theory of probability. It contains two measures 'plausibility' and 'evidence', which are allowing to model hints for supporting and contradicting a hypothesis separately. This leads to a measure for the ignorance, which does not exist within the conventional probability theory. The fuzzy theory can be derived from the kind of uncertain reasoning of the evidence theory. Therefore it is necessary to mention this approach within the context described here. The advantage lies in the simultaneous handling of different kinds of inputs like uncertain values, user experience inputs, measurement values etc. But the computerized implementation has to be designed dependent on the application.

Table 3.2 summarizes the described extended and alternative methods of this section and evaluates their advantages and disadvantages for the practical application.

In conclusion, a couple of methods are available to store and process knowledge of different kinds. These methods can be applied within all the layers of the autonomous component and on all levels of the autonomous system. These methods are unavoidable for the management and organization as well as the coordination to reach intelligent and autonomous behavior. This spreads widely from simple rules for taking decisions to optimization methods for the automated generation of optimized action schemes.

All these methods are basically focusing on the implementation of one single autonomous component operating within the environment of the autonomous systems. The interaction of several components leading to an optimized and self-organized behavior of larger parts of the autonomous system can be handled with agent technologies, which are explained in the following section.

Table 3.2. Characteristics of extended and alternative methods

Kind of Knowledge	Method	Advantages	Disadvantages
data and process knowledge based, supervised (input and output datasets and process relations)	Neuro-Fuzzy-Systems	- see Table 3.1 for Fuzzy and Percep-trons - validation and tuning implemented in the method	- see Table 3.1 for Fuzzy and Percep-trons
data based, supervised and optimization	Evolutionary Neural Networks	- see Table 3.1 for Perceptrons - topology optimization included in method	- see Table 3.1 for Perceptrons - very high training effort
data based self-organized (input datasets)	Self-Learning Self-Organizing Maps	- see Table 3.1 for SOM - adaptation to varying input data	- see Table 3.1 for SOM - adaptation hard to parameterize
	Growing Neural Gas	- adaptation to varying input data	- adaptation hard to parameterize - strongly dependent on dimensionality of problem
all kinds of knowledge	Evidence Theory	- very flexible applicability	- reasoning scheme must be designed for each problem appropriately

3.4 Agent Technologies

The autonomous system architecture introduced in chapter 1 and 2 denotes, that agents act autonomously on the base of information from the environment or other agents. In the simplest form the coordination layer requests or collects this information. If some information is missing the agent substitutes autonomously its original action scheme by a new one without this information. This behavior can be called intelligent and the necessary methods are introduced above.

Beyond this behavior of single agents, teams of agents shall be enabled to achieve a common goal. For example protection devices of power systems have the common goal to keep the system running by switching off faulty parts. The protection devices are operating on the same hierarchical level and are instances of the same type. Each device acts as a local agent but the goal is globally and can usually only be achieved by the action of more than one protection device. For all kinds of functionalities where a task can only be achieved by groups of agents special interaction schemes and methods are necessary.

Two basic configurations of agents have to be distinguished - a group of agents with different subtasks and groups of agents of the same kind and on the same hierarchical level. In both configurations the groups of agents fulfill a common goal. The first configuration is covered by the autonomous system definition in which autonomous components as agents request and provide information to other agents. The second configuration needs particular methods like for example negotiation to reach global goals while keeping local constraints. These methods or interaction schemes of agents have to be defined with respect to communication and coordination of the agents. The theory of intelligent agents [34][35] proposes several of these interaction schemes, which will be introduced shortly in the following. Further specific methods are explained together with their applications in the application chapters.

3.4.1 Communication

Communication protocols enable agents to exchange and understand messages. Interaction protocols enable conversation through structured exchanges of messages. A communication protocol needs a well-specified set of messages, which can be exchanged between two agents. An example for such a set of messages is the following:

- Propose a course of action
- Accept a course of action
- Reject a course of action
- Retract a course of action
- Disagree with a proposed course of action
- Counterpropose a course of action

A communication protocol is specified by a data structure with for example the following fields:

- Sender
- Receiver
- Language in the protocol
- Encoding and decoding functions
- Actions to be taken by the receiver

An example of such a communication protocol is the knowledge query and manipulation language (KQML), which is the proposal for a standardized agent communication [36].

3.4.2 Coordination

Coordination methods serve to establish conversation between agents. The coordination is based on the two components communication and individual decision-making. Different kinds of coordination are shown in figure 3.2 [36].

In general, cooperation is coordination among non-antagonistic agents, while negotiation is coordination among competitive or simply self-interested agents. Typically, to cooperate successfully, each agent must maintain a model of the other agents, and also develop a model of future interactions.

Cooperation and negotiation serve to reach a global optimum by individual actions. The global goal can be described by a value function for multi-objective decisions. Such a value function can be extended to a utility function to accommodate uncertainties in the consequence of an action. Via such a utility function the agents may reach an optimal or coherent behavior.

In cases where agents have conflicting goals or are simply self-interested, the objective of the coordination scheme is to maximize the payoffs (utilities) of the agents.

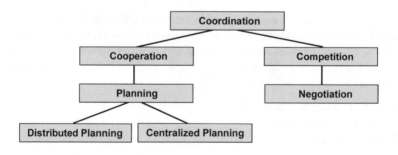

Fig. 3.2. Coordination of agents

In cases where the agents have similar goals or common problems, as in distributed problem solving, the objective of the coordination schemes is to maintain globally coherent performance of the agents without violating autonomy through, for example, explicit goal control.

Cooperation

A basic strategy for cooperation is the decomposition and distribution of tasks. The following mechanisms are used to distribute tasks:

- Contract net: announce, bid and award cycles
- Market mechanism: tasks are matched to agents by generalized agreements or mutual selection
- Multi-agent planning: planning agents have the responsibility for task assignment

In the contract net an agent wanting a task solved (manager) receives bids from agents being able to solve the task (contractors). Each potential contractor evaluates task announcements to determine if it is eligible to offer a bid. The contractor chooses the most attractive task and offers a bid to the corresponding manager. A manager receives and evaluates bids for each task announcement. The contractor receives an announced award message. If no bids are offered due to busy contractors or other tasks of more interest, the manager can adapt its task announcement. In this set-up the contractors act autonomously to a certain extend despite of the manager instance on a higher hierarchical level. The market mechanism and multi-agent planning are as well using a higher instance for the cooperation and planning.

Negotiation

In contrast to the cooperation scheme, the competition uses negotiation to reach a joint decision by two or more agents. This mechanism does not require a central decision maker or managing instance. The agents have to exchange their actual status or positions and then try to solve possible conflicts by themselves. The agents need a negotiation scheme and a decision process that each agent uses to determine its positions, concessions, and criteria for agreement.

An agent has to formulate the condition of its satisfaction related to actions to be negotiated. For example, an agent forms and maintains its commitment to achieve a common goal individually if it has not pre-committed itself to another agent to adopt and achieve the common goal, if the common goal is not contradictory to the individual goal and it is willing to achieve the common goal [37].

If we apply this negotiation scheme to a number of local voltage controllers of a power system the common goal is to keep all voltages within a given interval. The individual goal is to keep the local voltage as close as possible to the nominal voltage. A local controller is willing to contribute to the common goal by adding it to its individual goal. Each controller as an agent is following the local goal. If one agent is not able to reach the local goal, other agents may contribute by leaving slightly the local goal and contributing to the common goal. The negotiation optimizes between the common and the local goal.

Another way of describing the negotiation is based on the assumption that the agents are economically rational [38]. Agents are creating a deal that is a joint plan between the agents that would satisfy all of their goals. The utility of a deal for an agent is the amount it is willing to pay minus the cost of the deal. Each agent wants to maximize its own utility. The agents discuss a negotiation set, which is the set of all deals that have a positive utility for every agent.

These negotiation schemes as well as the cooperation schemes before need to be formulated and adapted in detail for the specific application area. These adaptations together with detailed descriptions will be given in the following application chapters. The methods of computational intelligence together with the agent technologies are serving to enable the characteristics of autonomous systems.

References

[1]   Poole D, Mackworth A, Goebel R (1998) Computational Intelligence: a logical approach. Oxford University Press Inc., New York

[2]   McDonald JR, McArthur S, Burt G, Zielinski J (1997) Intelligent Knowledge Based Systems in Electrical Power Engineering. Kluwer Academic Publishers, Boston

[3]   Madan S, Bollinger KE (1997) Applications of artificial intelligence in power systems. Electric Power Systems Research, Elsevier, no 41, pp 117-131

[4]   Dahhaghchi I, Christie RD, Rosenwald GW, Liu CC (1997) AI application areas in power systems. IEEE Intelligent Systems & Their Applications, vol 12, no 1, pp 58-66

[5]   Liu CC, Pierce DA, Song H (1997) Intelligent system applications to power systems. IEEE Computer Applications in Power, vol 10, no 4, pp 21-24

[6]   Bann J, Irisarri G, Kirschen D, Miller B, Mokhtari S (1996) Integration of Artificial Intelligence Applications in the EMS: Issues and Solutions. IEEE Transactions on Power Systems, vol 11, no 1, pp 475-482

[7]   Dillon TS, Niebur D (1996) Neural Networks Applications in Power Systems. CRL Publishing, Leics

[8]   Lai LL (1998) Intelligent system applications in power engineering: evolutionary programming and neural networks. John Wiley & Sons, Chichester

[9]   King RL (1998) Artificial neural networks and computational intelligence. IEEE Computer Applications in Power, vol 11, no 4, pp 14-16

[10]   Aggarwal R, Song Y (1997) Artificial neural networks in power systems. I. General introduction to neural computing. Power Engineering Journal, IEE, vol 11, no 3, pp 129-134

[11]   Aggarwal R, Song Y (1998) Artificial neural networks in power systems. II. Types of artificial neural networks. Power Engineering Journal, IEE, vol 12, no 1, pp 41-47

[12]   Aggarwal R, Song Y (1998) Artificial neural networks in power systems. III. Examples of applications in power systems. Power Engineering Journal, IEE, vol 12, no 6, pp 279-287

[13]   Wehenkel LA (1998) Automatic Learning Techniques in Power Systems. Kluwer Academic Publishers, Dordrecht

[14] Kohonen T (1984) Self-Organization and Assoziative Memory. Springer-Verlag, Berlin

[15] Kohonen T (1995) Self-Organizing Maps. Springer-Verlag, Berlin

[16] Song Y, Johns AT (1997) Applications of fuzzy logic in power systems. I. General introduction to fuzzy logic. Power Engineering Journal, IEE, vol 11, no 5, pp 219-222

[17] Song Y, Johns AT(1998) Application of fuzzy logic in power systems. II. Comparison and integration with expert systems, neural networks and genetic algorithms. Power Engineering Journal, IEE, vol 12, no 4, pp 185-190

[18] Song Y, Johns AT (1999) Applications of fuzzy logic in power systems. III. Example applications. Power Engineering Journal, IEE, vol 13, no 2, pp 97-103

[19] Sarfi RJ, Salama MMA, Chikhani AY (1996) Applications of fuzzy sets theory in power systems planning and operation: a critical review to assist in implementation. Electric Power Systems Research, Elsevier, vol 39, no 2, pp 89-101

[20] Srinivasan D, Liew AC, Chang CS (1995) Applications of fuzzy systems in power systems. Electric Power Systems Research, Elsevier, vol 35, no 1, pp 39-43

[21] Momoh JA, Ma XW, Tomsovic K (1995) Overview and literature survey of fuzzy set theory in power systems. IEEE Transactions on Power Systems, vol 10, no 3, pp 1676-1690

[22] Dillon TS, Laughton MA (1990) Expert System Applications in Power Systems. Prentice Hall International

[23] Germond A, Fink L, Bretthauer G, Wollenberg BF, Liu CC (1993) Exploring user requirements of expert systems in power system operation and control. Electra, no 146, pp 68-83

[24] Bretthauer G, Handschin E, Hoffmann W (1992) Expert systems application to power systems-state-of-the-art and future trends. Control of Power Plants and Power Systems, Selected Papers from the IFAC Symposium, Pergamon, Oxford, pp 463-468

[25] Goldberg DE (1989) Genetic algorithms in search, optimization and machine learning. Addison Wesley

[26] Schwefel HP (1995) Evolutionary and Optimum Seeking. John Wiley & Sons, Inc., New York

[27] Miranda V, Srinivasan D, Proenca LM (1998) Evolutionary computation in power systems. International Journal of Electrical Power & Energy Systems, Elsevier, vol 20, no 2, pp 89-98

[28] Yao X (1999) Evolving Artificial Neural Networks. IEEE Proceedings, vol 87, no 9

[29] Kiendl H (1998) Self-Organizing Adaptive Moment-Based Clustering. IEEE International Conference on Fuzzy Systems, Anchorage, Alaska, IEEE Press, Piscataway, NJ, vol 2, pp 1470-1475

[30] Martinetz T, Schulten K (1991) A 'neural gas' network learns topologies. in Kohonen T, et al, Artificial Neural Networks, North Holland, Amsterdam, vol I, pp 397-402

[31] Fritzke B (1995) A growing neural gas network learns topologies. Advances in Neural Information Processing Systems 7, MIT Press, Cambridge

[32] Shafer G (1976) A mathematical theory of evidence. Princeton Univ. Press, London

[33] Dempster A (1967) Upper and lower probabilities induced by a multivalued mapping. Annals of Mathematical Statistics, vol 38, pp 325-339

[34] Woolrigde M (2002) Introduction to Multiagent Systems. John Wiley & Sons

[35] Weiss G (2000) Mutliagent Systems: A Modern Approach to Distributed Artificial Intelligence. MIT Press

[36] Stephens LM, Huhns MN (2000) Multiagent Systems and Societies of Agents. in Weiss G, Mutliagent Systems: A Modern Approach to Distributed Artificial Intelligence. MIT Press, pp 79-120

[37] Haddadi A (1995) Towards a Pragmatic Theory of Interactions. Proceeding of International Conference on Multiagent Systems, San Francisco

[38] Rosenschein JS, Zlotkin G (1994) Designing Conventions for Automated Negotiation. AI Magazine, pp 29-46

4 Multi-Agent Negotiation Models for Power System Applications

James D. McCalley, Zhong Zhang, Vijayanand Vishwanathan, Vasant Honavar

Iowa State University, USA

Motivated by the need to coordinate decisions among competing power system stakeholders, we describe multi-agent negotiation systems to perform negotiated decision-making. The negotiation theory is reviewed. Two multi-agent negotiation models are described. A multi-agent negotiation system is implemented and some illustrative power industry applications are presented.

4.1 Introduction

There are a wide range of power system decision problems, traditionally falling under one of the three categories of operations, maintenance, and planning, with the delineation between categories derived from the nature of the decision and the time horizon. Some of these decision problems include generation dispatching, fuel scheduling, control-room preventive and corrective actions, incident restoration, transmission service scheduling, unit commitment, transmission equipment maintenance, control system planning, and transmission upgrades. All of these share common attributes, among which are:

- Resulting decisions have system-wide impact and therefore require significant coordination of information among the various decision-makers
- Higher system integrity is only achieved with greater allocation of financial resources
- The essential decision variables are inherently uncertain
- Uncertainty is reduced via acquisition and processing of information
- Important information tends to be spatially dispersed
- Complex and computationally intensive applications are required

As a result, decision-making support aids require modeling of multiple objectives, application of significant computational resources, and use of flexible data access capability. Mathematical programming has been and continues to be a mainstay for such decision problems. However, traditional optimization tools, by their very nature, assume the existence of a single and benevolent decision-maker that has centralized access to all information, and coordination between decision-makers is embedded in the processes and is generally of no threat to individuals carrying out those processes. Such was the case in the traditionally regulated world where ownership of all facilities, access to all information, and all authority for decision rested within the single umbrella of the vertically integrated utility company. With the advent of industry restructuring and associated organizational disaggregation, however, facility ownership is heavily fragmented, and information access and decision-making authority is quite limited for any one particular organization. Even more, the various organizations comprising the industry are not necessarily cooperative one with another; in fact, many portions of the restructured industry are intentionally organized to be competitive. Yet the need for coordinated decision remains, as it is essential to the operational integrity of the system. This necessitates a new paradigm to build upon and ultimately replace the centralized decision approach, enabling optimized decisions in an environment of highly distributed information and a multiplicity of competing stakeholders.

The industry has attempted to retain decision-making ability using traditional optimization tools, but it has come at the expense of forming new, centralized and competitively neutral authorities such as independent system operators (ISOs) and reliability authorities (RAs) to coordinate system operations and issues related to system reliability. These organizations arbitrate those decisions where conflict between two or more parties may otherwise arise. For example, during operationally stressed conditions having excessive risk of load interruption, a centralized authority generally selects the units to redispatch in order to mitigate the risk. Another example is maintenance: given requests for simultaneous maintenance outage of multiple components (generators, lines, and/or transformers) such that network integrity is excessively compromised, a centralized authority generally determines the sequence and timing of the maintenance tasks. In both of these cases, a conflict (which generators to redispatch in case 1 and which maintenance tasks to postpone in case 2) is settled by the arbitration of the central authority. The technology which motivates this book, multi-agent systems (MAS), may offer a viable alternative to this arrangement, or at least a useful complement, through the use of software agents equipped with negotiated decision-making capabilities operating within a MAS so as to coordinate decision-making of competing stakeholders. In multi-agent negotiation systems, stakeholders, represented by agents, engage in negotiation, proposing and counter-proposing until an outcome is identified that is satisfactory to all.

It is important to clarify terms at the beginning. A *stakeholder* represents a human individual or human organization that has interest in the stated decision-making problem would be one of the decision-makers if allowed to participate. Later, we use the term *player* and *party* to more explicitly refer to a stakeholder involved in a human-to-human negotiation. On the other hand, an *agent* is first and foremost a software entity, second, one that satisfies the usual criteria for agency (e.g., a computer system situated in some environment capable of autonomous action to meet its design objectives [1]), and third, for our purposes here, one that has the essential social ability of communicating (sending and receiving messages). Thus, a stakeholder may be represented by a party, or by an agent. A party always represents a stakeholder. Although an agent does not necessarily have to represent a stakeholder (there may exist "functional" agents, for example), we assume in this chapter that an agent does. In effect then, within the domain of this chapter, "agent" is the software encapsulation of "party."

Here, a software agent, armed with a coded negotiation model, represents each stakeholder, and conflict resolution is achieved via inter-agent message exchange until agreement is reached. MAS is an essential enabling technology because it provides the necessary infrastructure in terms of model instantiation and maintenance together with the communication needs, including messaging, directory services, and communication protocols [1].

There are at least two distinct ways in which the power industry will benefit from successful implementation of multi-agent negotiation systems: better decisions and better models. Better decisions may be expected because: (a) The ability to perform computer-evaluation of relevant issues, including accessing complex, distributed data, is inherent to the negotiation itself. This is in contrast to human-based negotiation where obtaining additional information or performing additional processing is typically done outside the time and space given to the negotiation process. (b) The negotiation speed is significantly increased. In contrast to human-based negotiation where negotiation speed depends on the limitations of the human negotiators, computer-based negotiated decisions may be reached as fast as network bandwidth and computer processing power allow. This not only provides for enhancing existing negotiated decision-making scenarios but also introducing negotiated decision-making where it was previously thought to be untenable. For example, networked negotiation enables consideration of negotiated decision-making between control centers, even between individual generation and/or transmission companies, following outages when typically decision-time is quite short.

The second way in which the power industry will benefit is that multi-agent negotiation systems offers a modeling framework that enables study of important power industry characteristics for which good models are not presently available. One of these characteristics is the level of centralization in decision-making, i.e., negotiated decision-making allows simulation and study of decision quality as the decision framework moves from being highly centralized (i.e., vertically integrated utilities) to entirely distributed. Another characteristic that can be studied with multi-agent negotiation systems is the flow of information; in particular, one can document the movement of information as well as the amount. Such documentation offers to illuminate organization effectiveness as a function of information flow, and to study the effect of providing new information flow paths.

The rest of this chapter is organized as follows. Section 4.2 reviews literature in terms of (a) negotiation theory and (b) computer-based negotiations. Section 4.3 describes implementation issues together with a description of the infrastructure we have used in our work. Section 4.4 describes power-industry application areas. Section 4.5 concludes.

4.2 Negotiation Theory and Agents: A Review

Endowing agents with advanced social abilities, such as negotiation, for use within multi-agent systems, has been of interest since the 1980's, but study of negotiation as a fundamental form of human interaction has been ongoing throughout the 20th century, and an awareness of these developments is essential for understanding the recent and intimately related work in multi-agent negotiation systems. Section 4.2.1 focuses on literature from decision science, economics, and anthropology; more recent literature related to computer-based negotiation mechanisms is discussed in Section 4.2.2.

4.2.1 Basics of Negotiation Theory

Negotiation is a fundamental form of human interaction, and we see it in labor-management disputes, international diplomacy, governmental processes, business relations, and interpersonal relations. Despite its prevalence throughout all human history, it was not until the middle of the 20th century before development of a theory for negotiation was initiated. This effort had roots in a number of different disciplines, including decision science, economic bargaining theory, social psychology, political science, industrial sociology, and social anthropology [2]. We do not attempt a comprehensive literature review here but rather provide basic concepts on which we draw in instantiating negotiation models in agents.

There are two sub-disciplines within decision science that need particular attention in order to do justice to the field of negotiation theory. The first is multi-criteria decision-making (MCDM), because it is MCDM that provides a number of different decision approaches for multi-criteria decision problems. Some of these approaches include [3][4] weighting methods such as analytical hierarchy process, Electre IV, goal programming, evidential theory, and utility-based approaches, where we typically search for efficient solutions, i.e., those solutions for which there exist no other solutions that can outperform them in all criteria. Of the various MCDM approaches, it is the utility-based approaches that have had particular influence on evolution of negotiation theory. A well-known and simple decision criterion is to choose the action, which maximizes the expected value of benefit. Thus, if we can associate with each course of action A a set of outcomes characterized by their benefits c_1, c_2, ...c_n and corresponding probabilities p_1, p_2,...p_n, we desire to select the course of action that has the largest value of $\sum p_i c_i$. Bernoulli [5], and later von Neumann and Morgenstern [6] and others [7-10] argued that rather than using expected value, the rational way for people to evaluate decision problems is on the basis of expected utility $EU(A)=\sum p_i u(c_i)$ where $u(\bullet)$ is a utility function that characterizes the decision-makers preferences with respect to the possible benefits of each outcome.

The second sub-discipline that needs particular attention is the theory of *competitive problems*, characterized by decision scenarios where certain of the decision variables are controlled by two or more independent parties having different interests. This discipline, which has largely grown out of utility-based decision approaches, has formed the basis for much of the non-agent and agent-related work in negotiation theory. The most influential aspects of this work fall under the *theory of games* in which two or more *players* (i.e. parties) choose courses of action and in which the outcome is affected by the combination of choices taken collectively [2][6-11]. A key assumption is that players behave *rationally*, where rational behavior is characterized by action selection, by each party, so as to maximize individual expected utility. Additional assumptions include (a) there is a fixed set of rules that specify what courses of action can be chosen; (b) there are well-defined end-states that terminate the game; (c) associated with each end-state are player-specific payoffs; (d) all players have perfect knowledge with regard to the rules, the range of outcomes, probabilities, and payoffs, and each player's preferences; (e) there is no interference or influence from the outside world. The central question addressed is: for a specified game, under assumptions a-e, what will be the utility vector on which the players will agree? The most well known example of such a game is the so-called prisoners' dilemma whereby the district attorney has two robbers in different cells. If both confess, both get 8 years jail time; if neither confesses, both get 5 years, and if one confesses and the other does not, the confessor gets 2 years and the other 10. Game theory provides several different models for studying player decisions.

A simple game-theoretic model is identified by Raiffa in [12], where it is assumed that by analyzing the consequences of no agreement, each party can establish a threshold value to be used for decision. Define x^* as the final-contract value, the sellers reservation price s represents the minimum price for which the seller will sell, the buyers reservation price b represents the very maximum price for which the buyer will buy. Then the zone of agreement is the interval (s, b), assuming $s<b$. If $b<s$, then agreement is not possible. The buyer's surplus is $b-x^*$, the sellers surplus is x^*-s, and both buyer and seller try to maximize their surplus.

Other game theoretic models pertaining to negotiation are identified by Cross in chapter 2 of [13], the most important of which is Zeuthen's two-party model. Here, if we define x_0 as the demand offered by party 2, party 1 identifies its expected payoff utility from demanding x according to $\Delta u_1 = u_1(x)[1-p(x)]-u_1(x_0)$, where $p(x)$ is the probability that party 1's insistence on x will result in a walk-out by party 2 (and $1-p(x)$ is the probability that party 1 will agree). If $p(x)$ is too large, then Δu_1 could be negative implying expected utility decreases for party 1, an undesirable situation. Therefore, in order to decide whether to counter-propose with x, party 1 determines the highest $p(x_m)$ that does not give $\Delta u_1<0$. This is found from $\Delta u_1 = 0 = u_1(x_m)[1-p(x_m)]-u_1(x_0)$, resulting in $p(x_m)=[u_1(x_m)-u_1(x_0)]/u_1(x_m)$. Thus, party 1 should counter-propose a certain value x only if $p(x)<p(x_m)$. Party 2 is assumed to use a similar model.

Cross [13] also provides a more advanced two-party model where time and money are the issues. Each party knows their own desired settlement, q_1 and q_2, respectively. Also, each party has an expectation of their opponent's rate of concession (change in offer per time period); party 1 uses r_2 and party 2 uses r_1. Then there is a cost to each party for each time interval that passes without agreement, C_1 for party 1 and C_2 for party 2. There is also a discount factor to express the money in terms of present value, so that the present value cost is given by $Z=(C_1/a)(1-e^{-aT})$ where a is the discount factor and T is the expected time to reach agreement given by $T=(q_1+q_2-M)/r_2$. Then the total present value to party 1 of targeting a desired settlement is given by $U_1=u_1(q_1)-Z$. Cross identifies increasingly rigorous ways to update r_2 (learning) based on the negotiation process.

One limitation to game theory is that it is pre-occupied with outcome, discussed in terms of equilibria (e.g., Nash, perfect, dominant), rather than the process (or mechanism) used to arrive at that outcome. This point is central to the goal of automating multi-party decision-making because we need the capability of implementing the mechanism to achieve the goal. Thus, we turn to the closely related negotiation theory.

There are at least six different mechanisms of reaching a collective decision among 2 or more parties, including persuading, educating, manipulating, coercing, appealing to an authority, and negotiation [14]. Of these, the last two, arbitration and negotiation, are two distinctive forms of identifying agreements between two or more parties that have significantly more formality and structure. Arbitration provides a mechanism, which selects a single outcome as the point of agreement between the parties. Judges often assume this role in legal disputes, and, as observed in Section 4.1, so do certain kinds of power system decision-making authorities. In arbitration, the parties direct their communication towards a third party, but not to each other. Arbitration, with only a single decision-maker, is most effective when parties seek to agree over values, norms, and the assessment of facts. Negotiation, on the other hand, is the *joint-decision* process of forming and revising offers by each involved party, whereby offers are made with the intention to converge to an agreement, without the presence of a third-party decision-maker. (Reference [2] argues that this definition applies more appropriately to *bargaining* with a broader definition used for negotiation that includes the initiation and recognition of the motivating need, the process, the final outcome, and the execution of that outcome). In negotiation, the focus of communication is (are) the other party (parties). Generally, negotiation is performed as a result of a conflict or dispute between two or more parties, and the negotiation objective is to resolve the dispute. Negotiation is most effective in a situation of scarcity when parties seek the same resources without there being enough to satisfy both [2]. These types of negotiations have been characterized as either *strident antagonist* or *cooperative antagonist* [12]. The former is characterized by completely distrustful and malevolent (towards one another) parties, as would be the case when authorities negotiate with kidnappers or airline hijackers. The latter is characterized by entirely self-interested and disputing parties but ones that recognize and abide by whatever rules exist. A third type of negotiation is called *fully cooperative* [12], where the parties have different needs, values, and opinions, but they share information, expect total honesty, perform no strategic posturing, and think of themselves as a cohesive entity with intention to arrive at the best decision for the entity, as would be the case for a happily married couple. We are interested here in the two different levels of cooperative negotiations, since they better typify the various types of power system decisions problems. For example, a negotiation involving two transmission owners over equipment maintenance schedules is a good example of cooperative antagonists. A negotiation involving two ISOs sharing responsibility for operational integrity of different portions of the network in the same interconnection, over equipment maintenance schedules, is a good example of a fully cooperative negotiation. Some negotiations are also characterized by the presence of a mediator, where an impartial outsider has the role of helping the parties find a compromise solution. Although often useful, this refinement offers no fundamental change to negotiation models, and we do not address it further.

There are at least two main phases to any negotiation. These are:

1. Information exchange:
   - Pre-bargaining, including identification of the issues, establishing maximal limits to the issues, and agreeing on the rules
   - Issue iterations (offers and counteroffers)

2. Arriving at the outcome:
   - convergence on a final contract value
   - retention of the status quo (no change) via a walk-out by one party

Characterizing features of negotiations have been set forth in a number of works, including [2,12,13,15]. Of these, a key attribute is whether the negotiation involves 2 parties (bilateral) or more (multi-lateral). Multi-party negotiations have complexity that significantly exceeds that of bilateral negotiations, as players may form any of a number of different coalitions. Even in the simplest of cases, the three-party negotiation (A, B, C), one must account for any of four scenarios: no coalition, or coalition of AB, AC, or BC. One obvious approach is to abandon negotiation altogether and utilize, for example, voting, some form of arbitration such as an auction, or perform the multi-party negotiation as a sequence of independent bilateral negotiations. Another approach, which maintains the essence of the multi-party negotiation, suggested in [16] and described further in [17], called Rubenstein's model of alternating offers, formulates the negotiation rules to explicitly disallow coalitions. Here, when one of the agents makes an offer, all other agents respond, with each agent accepting, rejecting, or canceling (walking out). The negotiation terminates if all agents accept the offer (an agreement), or if one of them cancels. If the negotiation does not terminate, the negotiation proceeds to the next time period, another agent makes an offer, and the process repeats. Coalitions are prevented by disallowing inter-party communication. In addition, it is important to ensure that no party knows others' responses until the round is complete (otherwise parties have incentive to wait, to gain more information). The order in which parties can offer is randomized. In addition to avoiding coalitions, Rubenstein's model is also attractive because it is general; it works just as well for bilateral negotiations as it does for multi-lateral negotiations.

A second key attribute is the number of issues over which the negotiation occurs; there may be 1 (single issue) or more (multi-issue). The ability to handle multi-issue negotiations lies in the way each party evaluates an offer. Some method of normalizing among different issues is required, and expected utility provides for this. Other features of most negotiation models are whether or not they represent the influence of time, learning, strategic behavior, and a pre-bargaining phase.

In reading the negotiation literature, it is important to recognize whether the author's perspective is *descriptive* or *prescriptive*. Descriptive models describe how negotiating parties actually behave, whereas prescriptive models prescribe how the parties should behave. For example, Gulliver in his well-known text [2] argues strongly against the use of utility theory in negotiation models because, he feels, it does not represent how people actually think (no one, he argues, actually decides based on quantification of their own and others' probabilities and preferences for various outcomes) and instead proposes two other models that capture the *cycling* and *developmental* features of negotiation, respectively. Considering Gulliver's criticism in the context of multi-agent negotiation systems, one may argue that it provides a degree of information accessibility and modeling power that was not available when Gulliver wrote, so that many of the complexities of how humans decide may now be effectively addressed. However true, this response neglects to recognize that multi-agent negotiation systems are not concerned with describing how humans negotiate (although it may be useful to build into multi-agent negotiation systems certain features of how humans negotiate). Rather, multi-agent negotiation systems are concerned with providing humans with decision-support, i.e., prescribing how humans should decide. Therefore, the implementation mechanism is important only insofar as it provides us with desirable outcomes. In this sense, then, a multi-agent negotiation system has the same function that mathematical programming has had, with the distinct advantage that the former accommodates the feature of distributed decision-making that is now prevalent in the electric power industry.

4.2.2 Computer-Based Negotiation Systems

There is a large and growing body of literature on computer-based negotiation systems. The most important of these include texts published in 1994 [18], 1999 [19], and 2001 [17] that well-capture the state of the art at those times. In addition, references [20-25] represent the important recent efforts in the area.

Rosenschein and Zlotkin [18] make a strong case that "game theory is the right tool in the right place for the design of automated interactions," arguing that despite its shortcomings in capturing human interactions, automated societies are perfectly amenable to the assumptions on which game theoretic models rest. They provide a list of attributes associated with machine interaction, including efficiency (outcomes should be Pareto Optimal), stability (no agent should have incentive to deviate from the agreed-upon available strategies), simplicity (low computational and communication requirements), distributedness (interaction rules should not require a centralized entity), and symmetry (the interaction mechanism should not arbitrarily favor one agent more than another). The work rests on the standard assumptions of game theory (rationality based on utility) together with a few more reminiscent of Rubenstein's model, including: (a) each negotiation is independent of past or future negotiations; (b) agent-specific utility calculations may be transformed into common "system" units; (c) all functionalities (abilities) are equally accessible to all agents; (d) public agreements are binding; (e) no utility (in the form of money, for example) is explicitly transferred from one agent to another. They make the important clarification that their work is about design of machine negotiation protocols, where protocols in this context pertain not to the low-level issue of how machines communicate (it is assumed that they do), but rather to a higher-level issue regarding the public rules by which machines come to agreement, such as Rubenstein's model described above. Thus, they proceed to identify different problem domains and specify various protocols appropriate for that domain. For example, the task-oriented domain is one in which an agent's activity can be defined in terms of a set of tasks that it has to achieve (in contrast to domains where agents are concerned with moving its environment from one state to another, or where agents assign a worth to states and select the best state in which to move). Given a protocol, the remaining attributes necessary to characterize a negotiation are the space of possible deals, the negotiation process, and the negotiation strategy. They utilize standard game theoretic models (e.g., Zeuthen's) to analyze the influence of different strategies on outcomes.

Huhns and Stephens [19] also emphasize the importance of protocols, and they distinguish between communication protocols (e.g., KQML, KIF) and "interaction" protocols. They classify the different interaction protocols into coordination, cooperation, contract net, blackboard systems, negotiation, and market mechanisms. Their overview of each provides a useful taxonomy for broadly understanding negotiation protocols.

Kraus [17] clearly distinguishes efforts in the area of designing agent *interaction* (i.e., coordination and cooperation) from that of designing agent *architecture*. Her efforts, relating to the former, integrates game theory with economic techniques and artificial intelligence heuristics to develop a strategic-negotiation model patterned after that of Rubenstein under assumptions similar to those of Rosenschein and Zlotkin. The importance of this work is in its detailed treatment of illustrating the generality of the proposed negotiation model in a diverse array of applications, including negotiations about data allocation, resource allocation, task distribution, pollution reduction, and hostage crises.

Reference [20] proposes a theory of negotiation developed with the intent of understanding negotiation support systems as computer-based decision systems, predicated on the idea that there are 8 different features that must be identified in order for a negotiation to be properly understood. These features are issues to be negotiated, entities involved, the acceptance region of the entities in the space of issues, the current location of the entities within that space, the strategies and movements of the entities, and negotiation rules, and the level and nature of assistance from an intervener (arbitrator or mediator). A long list of computer-based support functionalities is provided for each of the features, and a classification framework is provided in terms of the kind of entity set for which the system is used (group/peer-to-peer or organization/hierarchical) and the nature of the system's participation in the negotiation (assistance/support or autonomous negotiator). Reference [21] identifies 5 negotiation activities where mathematical modeling can provide prescriptive decision aid, and focuses on one of them, the search for agreement and improvements, in showing how it can be formalized as a MCDM gradient search problem or as a constraint proposal problem. Reference [22] identifies characterizing features of distributive and integrative negotiations, terms first articulated by [26] to distinguish between "fixed-pie" negotiations where parties are inherently in conflict and compete over scarce resources such that when one party gets more, the other gets less (distributed) and win-win negotiations where some settlements can be better for both parties.

Although integrative negotiations are generally multi-issued, they do not have to be as illustrated in the classic case where two sisters argue over an orange, one needing the juice and the other the peel. It is distributed as long as they do not know each others' needs but immediately becomes integrative when they do. Integrative negotiations generally lead to better solutions, and the authors conclude that auctions should be considered when distributive negotiation cannot be converted to integrative, but auctions are not applicable where it is possible for parties to learn about one another to determine opportunities and establish relationships. This theme is extended in [23] which purposes logrolling, an algorithmic method for multi-issued integrative negotiations that produces Pareto-optimal solutions through jointly improving exchange of issues such that loss in some issues is traded for gain in others resulting in overall gain for all parties. Reference [24], in focusing on computer-based business-to-business negotiations, also distinguishes between distributive and integrative negotiations and their relation to auction as in [22] and goes on to also compare norm and goal orientations in designing negotiation protocols. Reference [25] further addresses automated negotiation in the context of e-commerce applications, providing a useful negotiation taxonomy that collectively incorporates many of the attributes discussed piecemeal in the literature to date. Within this taxonomy, the parameters of the negotiation space include: cardinality (number of agents, number of issues), agent characteristics (role, rationality, knowledge, strategy, etc), environment (static or dynamic), goods (public or private), and parameters related to offers, information, and allocation. It also describes a number of proposed negotiation models and locates them in its taxonomy.

4.3 A Multi-Agent Negotiation System

We have developed a MAS and instantiated agents with negotiation capabilities in order to implement automated decision-making among competing stakeholders within a power system. Section 4.3.1 describes the MAS implementation; Section 4.3.2 and Section 4.3.3 describe the negotiation models that we have used. The system is used in Section 4.4 to facilitate decision problems related to line loading and maintenance, respectively.

4.3.1 MAS Implementation

In our work, we built a platform independent Java-based API called *MASPower* [27-29] on top of the commercial distributed computing platform Voyager ORB [30] to instantiate agents and multi-agent systems for facilitating coordinated and negotiated decision-making from power system decision-makers. Voyager supports dynamic proxy generation, naming services, synchronous and asynchronous messaging, management of multiple concurrent tasks and multiple conversation protocols, and preceptors for accessing local and remote percept sources for distributed MAS. In this framework, we used object-oriented software design methods to develop agents representing different power system entities, e.g., suppliers, transmission owners, system operators, and delivery companies. The developed software is organized into eight packages: basic and collaborative agent classes to implement agents with different functionalities, agent GUIs, tasks carried out by agents, functionalities for enabling inter-agent messaging, functionalities for enforcing conversation protocols, interfaces for directory services, interfaces for distributed computing inter-agent messaging, and classes for enabling interagent negotiations. Individual agents may reside on any CPU within a network as long as the CPU is running *MASPower* on top of Voyager.

We extended the federated directory service implementation of Voyager ORB to provide the ability to maintain names of currently active agents together with keywords to identify the agent's area of expertise. *MASPower* stores the directory location as an XML document, read by every newly created agent, to avoid the need to recompile a program every time the directory location is changed.

Agent communication is performed using inter-agent messaging with message interpretation being private to each agent, providing the ability to interpret the same message differently under different agent internal states. Structural elements of an inter-agent message are per FIPA-2000 recommendations [31] and include sender, receiver, content, ontology, conversation ID, protocol, reply-with, in-reply-to, and language. Multi-agent conversations are managed using thread, tagged by unique conversation identifiers generated by the agent initiating the conversation. Conversation protocols were designed as finite state machines (FSM) following the COOL notations [32]. The FSM for a conversation protocol is characterized by a START state, END state, FAIL state, and a variable number of intermediate states. Transition between one state to another occurs by either sending or receiving a message with a particular performative. For example, the FIPA recommendation for the contract net protocol [33,34] can be encoded as the FSM in Fig. 4.1. This protocol is useful for automated contracts in environments where all agents cooperatively work toward the same goal. The manager proposes a task, announces it, and potential contractors evaluate it (together with other announcements from other managers) and then submit bids on the tasks for which they are able to perform.

The FSM to be used by an agent depends on the role that the agent is playing in the conversation: the FSM in Fig. 4.1 is used by the agent responding to the initiating agent. The initiating agent uses the same FSM except that "send" and "receive" labels are interchanged for all transitions.

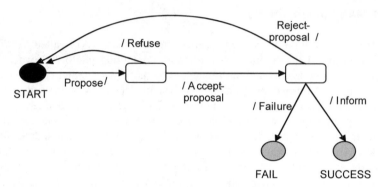

Fig. 4.1. FSM of Contract Net Protocol in COOL Notations

Each activity that can be undertaken by an agent in its lifetime is organized as tasks. Whenever a new task instance is created, the object registers with the agent's task manager. A key attribute of *MASPower* is that many tasks can run concurrently within a single agent.

4.3.2 Negotiation Protocol

We conceive of a society of agents organized as the electric power industry is organized, i.e., there are agents corresponding to load-serving entities, generation owners, transmission owners, and whatever centralized organizations that may exist such as ISOs, reliability authorities, and power exchanges. Decisions are made as a result of different inter-agent negotiations. Our negotiation protocol is bilateral, multi-issued, and integrative, and it may be applied to decisions with or without incorporation of uncertainty. We begin by describing the protocol in terms of multilateral negotiation, without uncertainty, as it is both general and simple. By avoiding the need to model uncertainty, agent decisions are made based on assessment of *utility* (or value), rather than *expected utility* (or expected value). One attractive feature of this paradigm is that successful termination of a negotiation is guaranteed to be *pareto optimal*.

An agent that initiates the negotiation process is termed *initiator*, and one or more responding agents are termed *responder(s)*. Our approach is patterned after that of Faratin, et al in [35] using value tradeoffs; the FSM of our protocol is illustrated in Fig. 4.2.

The agents negotiate over values for a set of *preferentially independent* issues (pp. 101 in [7]). When two issues are preferentially independent then each agent may assert its preferences for an increasing or decreasing value of one issue without any relation to the other issue.

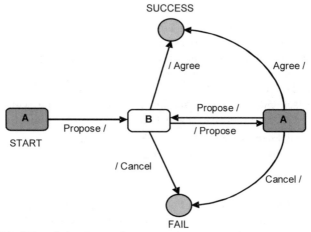

Fig. 4.2. FSM of Negotiation Protocol

When there are more than two issues, preferential independence is similarly defined for every subset of the set of issues and its complement. The set of issues for negotiations between any two entities usually revolve around some notion of {*quantity, quality, unit price*}. As proposed in [35], agents negotiate on the value of an issue that is within a delimited range, i.e., agents first mutually agree on a range of allowable values for every issue (pre-bargaining).

In a multi-agent system S consisting of at least two agents, an agent a wants to negotiate a value for a set of issues with agents in a subset of $S - \{a\}$; let $X_a = \{x_1, x_2, ..., x_n\}$ be the set of issues about which agent a wants to negotiate, each taking values in the range specified in the set:

$$range(X_a) = \{[\min(x_1), \max(x_1)], ..., [\min(x_n), \max(x_n)]\} \tag{4.1}$$

The set X_a is called a *negotiation set*. The agent uses a non-decreasing or non-increasing scoring function $V(x)$ to score the value of each issue between 0 and 1^1. Such functions are sufficient to model transitive preference structures. If the agent prefers an outcome x' to x'' for a single issue, then $V(x') > V(x'')$; if the agent is indifferent between two outcomes x' and x'', then $V(x') = V(x'')$. *MASPower* uses a model of additive value functions (p. 90 in [7]) to get the net value of a negotiation set, through what we call the scoring function of the negotiation set, as given by:

$$V^a(X) = \sum_{i=1}^{n} c_i * V(x_i) \tag{4.2}$$

---

[1] "x" is used to denote the negotiation issue and also the value for that issue whereas $V(x)$ denotes the score for that value of the issue. For example, an issue x can have a value of 50 when it is defined in the range [0,100] and a score of 0.6.

where the c_i are the relative importance of issues x_i to the agent, such that the weights are non-negative and sum to 1. Finding the relative importance of issues could be based on a subjective assessment by the human owner. This additive scoring function is the simplest multi-issued function with useful properties: two agents using additive scoring functions are sure to achieve a pareto optimal agreement, as shown in [12, p.164].

An agent uses either a non-decreasing or a non-increasing scoring function $V(x_i)$ to assign a score (between 0 and 1) to a permissible value of an issue x_i. *MASPower* implements scoring functions based on the implementation in [35]: For a non-decreasing issue x defined on the domain $[\min(x), \max(x)]$,

$$V(x) = (x - \min(x)/(\max(x) - \min(x))^k \qquad (4.3)$$

The family of curves is shown in Fig. 4.3. The curves have an interesting interpretation in case of utility functions: $0<k<1$ models risk-averse behavior. Risk-averse behavior is associated with a decision-maker who prefers an outcome with lower risk between outcomes with same expected consequences; $k>1$ models risk-prone behavior that is associated with a decision-maker who seeks risk; and $k=1$ models risk-neutral behavior (cf. pp. 148-157 in [7]). For an issue with non-increasing value function, the score obtained in Eq. (2) is subtracted from one.

If the agent wants to negotiate on an issue, the agent's value of this issue is modeled solely a function of its resources, i.e., the agent uses the current state of its resources to determine a value for an issue.

This is captured by the *negotiation functions*, which serve to generate a value of a single negotiation issue as a function of the agent's resources:

$$x = \sum_{i=1}^{n} w_i * f(r_i), \quad \sum_{i=1}^{n} w_i = 1 \text{ and } w_i \geq 0 \qquad (4.4)$$

where the r_i's are the "resources of the agent" to set the value for negotiation issues [35].

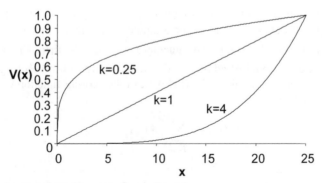

Fig. 4.3. Value Functions Corresponding to Eq. 4.3

An agent's resources are those quantities controlled by the agent that influence the agent's decision-making with respect to one or more negotiation issues. Agent resources could include, for example, money the agent has at its disposal (where it should be clear that a *resource* of money is distinct from a negotiation issue of money). Another typical resource is available negotiation time, since an agent typically knows a particular time by which a decision is required. A resource particular to our work is equipment life. One feature of this approach, where issue values are determined by agent resources, is that issue values, at any instant during the negotiation, do not depend on the offers that this agent receives from the other agents. An important implication of this is that the agent does not behave strategically in response to bids from other agents.

Although the function $f(r)$ can model complex dependencies, the current implementation of *MASPower* uses a family of parameterized functions for this purpose. The first step in generating the function $f(r)$ is to model the agent's individual interpretation of the significance of r. This is because for the same amount of resource (say time), some agents would prefer to concede rapidly when the time remaining to complete the negotiation is less, while others might prefer to hold on and not concede. If the normalized remaining resource is r (where $r = 1$ indicates all of the resource remains and $r = 0$ indicates none of the resource remains), an agent's individual interpretation of the significance of r is modeled as a function $g(r)$ using two parameters: x_0, the normalized initial value for an issue, and β, an index (cf. [35]), via the function:

$$g(r) = x_0 + (1 - x_0)r^{1/\beta} \qquad (4.5)$$

The agent's the negotiation function $f(r)$ for the issue x can be written as:

$$x = f(r) = x_{min} + g(r) * (x_{max} - x_{min}) \qquad (4.6)$$

for an issue x with increasing value function, and

$$x = f(r) = x_{min} + (1 - g(r)) * (x_{max} - x_{min}) \qquad (4.7)$$

for an issue x with decreasing value function.

Issues that depend on multiple resources are evaluated based on Eq. (4.4). The value for an issue that is prescribed by Eqs. (4.6) and (4.7) always fall within the allowed interval for the issue because of the parameterization. In other words, these equations never recommend an invalid value for an issue. The influence of parameter β, which models the attitude of the agent towards a resource, is best understood from Fig. 4.4.

When the agent has maximum resource available and as the quantity of the resource decreases (i.e., r changes from 1 to 0), a model with $\beta < 1$ concedes quickly, while there is still significant resource remaining, while a model with $\beta > 1$ concedes slowly, waiting for the resource to diminish further.

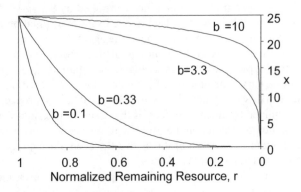

Fig. 4.4. Boulware and Conceder Negotiation Functions f(r)

As the resource diminishes, the latter model concedes rapidly while the former model concedes more slowly. The family of curves with $\beta > 1$ are called *Boulware* [2][12], whereas $\beta < 1$ are called *Conceder* [35].

The agent's decision-making algorithm under this negotiation paradigm is based on its score for the negotiation set. Suppose the agent a's private scoring function for a negotiation set X at the time instant t is $V^a(X)$. This score is purely based on the internal state of this agent and is a function of its value functions, tradeoffs, current state of resources, and negotiation functions. When this agent receives an inter-agent message containing an offer X' from another agent for the same negotiation thread at a time $t' > t$, agent a replies with either a refuse, accept, or counterproposal, as illustrated in Fig. 4.2.

4.3.3 Embedding Agents with Social Rationality

In this section, we develop an expected utility-based model for multi-agent rational decision-making, which is capable of dealing with uncertainty as well as enabling each agent to balance between its social behavior and self-interest.

As pointed out in [29], the value function based multi-agent negotiation model oversimplifies the preference and attitude of the agent and precludes the possibility of modeling uncertainty in the outcome of an action. However in practice, agents do face uncertainties in the consequences of their actions. While the expected utility theory has been widely adopted as the quintessential paradigm for decision-making in the face of uncertainty [9][10], the predominant view of an agent's decision-making function has been that it need be solipsistic[2] in nature, based upon the principle of maximizing the individual expected utility [36] given the probability of reaching a desired state and the desirability of that state. Al-

---

[2] Solipsistic (unlike *Social*) agents do not explicitly model other agents within a multi-agent system.

though this is intuitively and formally appealing, it lacks applicability in real systems consisting of multitude of interacting agents.

Given the importance of power system integrity, when designing a system in which multiple agents need to interact (coordinate) in order to achieve both *individual* (e.g. maximize individual entity's revenues) and *system* (e.g. minimize the system risk level) goals, we hold that neither egoism nor altruism are the best means to achieve globally optimal system states, but rather a good combination of these two aspects of interaction can yield the best global results. Because of the inherent interdependencies between agents, an agent's decision affects not only itself, but also other agents in the multi-agent system environment. It is therefore important to equip the agents within a multi-agent system with a mix of self-interest and social consciousness that allows them to value the performance of the entire society as well as their individual performance. An agent's decision should depend not only its individual utilities, but also on social utilities (utilities afforded to other agents in the MAS) of all possible actions while determining which action to perform. This is more appropriate for application to power systems where entity-represented agents are interdependent. If an agent places more emphasis on its individual utility, it is selfish in nature. On the other hand, an altruistic agent pays more attention to social utilities. A *socially rational agent* tries to maintain a balance between individual and social responsibilities [37].

Due to its intuitive and formal treatment of making decisions from a set of alternatives under uncertainty, we use the aforementioned expected utilities of the agent's actions to describe a dynamic utility calculating framework, which provides agents with a more descriptive notion of choice within a multi-agent environment. In our MAS, a particular agent may work in a group with a small number of agents, a loose confederation with a larger number of agents and hardly at all with the remaining agents. Thus, the calculation of social utility (includes individual utility) can be further distinguished by differentiating between the different social relationships in which an agent is engaged. To account for this, we define R as the agents' *Social Relationship Matrix* in our MAS:

$$R = \begin{bmatrix} r_{1,1} & r_{1,2} & \cdots & r_{1,n-1} & r_{1,n} \\ r_{2,1} & r_{2,2} & \cdots & r_{2,n-1} & r_{2,n} \\ \vdots & \vdots & \ddots & \vdots & \vdots \\ r_{n-1,1} & r_{n-1,2} & \cdots & r_{n-1,n-1} & r_{n-1,n} \\ r_{n,1} & r_{n,2} & \cdots & r_{n,n-1} & r_{n,n} \end{bmatrix} \tag{4.8}$$

R is a $n \times n$ matrix (n is the number of agents within the MAS); each row or column corresponds to an agent. Each $r_{i,j}$ should be a non-negative number; the value of $r_{i,j}$ indicates agent i's attitude toward the social relationship between itself and agent j. This means that each agent can quantitatively weight all social relationships with respect to the influence of its possible actions within the multi-agent system. While different agents may have different social perspectives, the value of $r_{i,j}$ is not necessarily equal to $r_{j,i}$. Therefore, in a general case the social relationship matrix R is not symmetric. This social relationship matrix could be maintained in a central secure repository[3]. Each agent can retrieve its own social relationship information from matrix R using its *Identification Matrix I*. This social relationship matrix coupled with individual identification matrix mechanism can be regarded as a security feature provided by the MAS platform similar to the private-public key distribution mechanism in computer system security. For instance, for the k-th agent, its identification matrix is a vector in the form of:

$$I = [i_1 \quad i_2 \quad \cdots \quad i_j \quad \cdots \quad i_n] \tag{4.9}$$

where $i_j = 1$, if and only if $j = k$; otherwise $i_j = 0$.

Supposing the k-th agent executes a possible action A, from its perspective, the utilities afforded to each agent in the MAS by A can be denoted as:

$$U_A = \begin{bmatrix} SU_{k,1} \\ \vdots \\ IU_{k,k} \\ \vdots \\ SU_{k,n} \end{bmatrix} \tag{4.10}$$

where $IU_{k,k}$ represents the individual utility to the k-th agent by that action, and all others are social utilities[4]. Then we can compute the expected utility for k-th agent if it carries out action A by the following equation:

$$EU(A) = I_k \times R \times U_A \tag{4.11}$$

By trying to maximize the above combined expected utility, agents can naturally take both individual and social utilities into the consideration of their decisions. The above mechanism equips the agents within multi-agent systems with a mix of self-interest and social consciousness that allows them to rationally evaluate their individual performance over the entire society. In addition, by varying the corresponding values in the relationship matrix R (if $r_{i,j}$ is set to zero, that means agent i either neglects the social relationship ($i \neq j$) between the two agents or totally removes its personal benefits from its decisions ($i = j$).), each agent can dynamically determine the way that it combines the individual and social utilities of all possible actions in order to make a rational decision.

---

[3] Neither the central repository nor other agents could have access to individual agent's social relationship information except that the agent is cooperative and willing to reveal its information.

[4] In a cooperative environment, when agent k executes a possible action A, we assume that each agent would let agent k know its utility offered by action A. This is simple and enables agents to avoid modeling other agents in the system.

4.4 Illustrations

4.4.1 Security-Related Decision-Making

We used the IEEE Reliability Test System [38] under stressed operating conditions, with the following decision required: *Determine by how much to operate a transmission circuit in excess of its identified rating?* In the traditional vertically integrated energy industry, this decision was made by a single organization, the utility company, because it both owned and operated the transmission system. However these two functions are now separated, with the ISO responsible for system operations and implementing the market based dispatch insofar as system security limits allow. A transmission owner is responsible for the physical integrity of the circuit, including the specification of the circuit rating, and in addition, the transmission owner receives revenues for use of the circuit in proportion to the flow. Implementations of common power system algorithms are made accessible to the agents from "C-code" implementations of the power system software, MATPOWER [39], and other codes. Data pertaining to the power system is maintained in a relational database to facilitate persistent storage and algebraic operations on it [40].

So we have simulated a negotiation between the ISO-Agent and the Transco-Agent over the increase in circuit rating and pro-rata compensation for the transmission service. Both agents employ the value-function based negotiation model discussed in section 4.3.2. The negotiation issues are *circuit rating increase* and *monetary compensation*. The resources used by the agents are *equipment life* (only for the Transco), *money*, and *negotiation time*. Fig. 4.5 illustrates the progress of the negotiation. The negotiation concluded after 54 iterations, taking 282 seconds, when the ISO accepts an offer of 4.62 MW rating increase at $13.13 for each MW of transmission service. A second simulation (not shown) repeated the first, except that the negotiation time resource for the Transco was decreased from 600 sec to 240 seconds, resulting in agreement after 42 iterations taking 225 seconds, at 5.54 MW rating increase at $11.76 for each MW of transmission service.

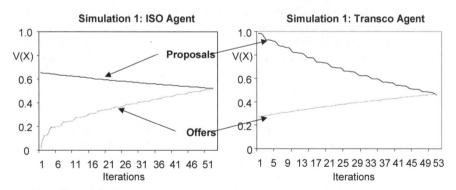

Fig. 4.5. Negotiation Process Between ISO Agent and Transco Agent

By decreasing the Transco's resource *'negotiation time'*, the Transco makes coarser changes in its proposals, thereby missing several good intermediate deals. We also completed another system security related inter-agent negotiation simulation using the rational negotiation model based on expected utility described in section 4.3.3. By applying several modifications to the original system [38], we constructed a security-constrained case with high system overload risk [41].

Representing the independent system operator, the ISO-Agent is in charge of the overall system security and periodically examines the system risk value calculated by RBSA-Agent[5] [42]. When it discovers high system overload risk, it immediately initiates negotiation with the Load-Agent representing the load entities at bus 13. The negotiation issues include *load curtailment* by the Load-Agent, and *compensating money* offered by the ISO-Agent. During the negotiation, the Load-Agent then has to analyze the tradeoff between the compensation proposed by ISO-Agent and the expected monetary loss due to its load curtailment.

The simulation results are shown in Fig. 4.6. The system overload risk significantly decreases as the Load-Agent agrees to shed more and more load at bus 13, and eventually when the two agents come to agreement that the Load-agent sheds 700MW load, the system overload risk drops to zero. The expected utilities of the two agents both mainly increase as the negotiation processes. This is because both agents use socially rational negotiation models: the system security level influences the ISO-Agent as well as the Load-Agent. Fig. 4.6 also illustrates how the compensated money evolves during the negotiation process, finally resulting in agreement between the two agents on $58,974 for 700MW load curtailment.

4.4.2 Maintenance Scheduling

There are thousands of high voltage power transformers in a bulk transmission system. Although power transformers have proven to be reliable in normal operation conditions with a global failure rate of 1~2% per year, the large investment in generating capacity after World War II, and continuing into the early 1970's has resulted in a large transformer population which now is fast approaching the end of life [43]. Therefore, with large number of aged electrical equipment in a system, certainly there would be numerous maintenance scheduling conflicts. For the sake of system security, it is important that priority be placed on scheduling system-wide maintenance activities to maximize the achieved overall risk reduction. Here we are interested in using multi-agent negotiation to solve this equipment maintenance-scheduling problem.

---

[5] *Risk Based Security Assessment* (RBSA), a probabilistic security assessment method, provides a quantitative risk index accounting for uncertainties in electric power system operating conditions, weather parameters, outage events, and equipment performance while condensing security assessment results into compact distributions used in economic decision-making paradigms. RBSA is implemented as a functional agent that provides system (or specified zone) risk and risk-sensitivity indices. These indices are computed to assess the system security levels as a function of existing and near-future network conditions.

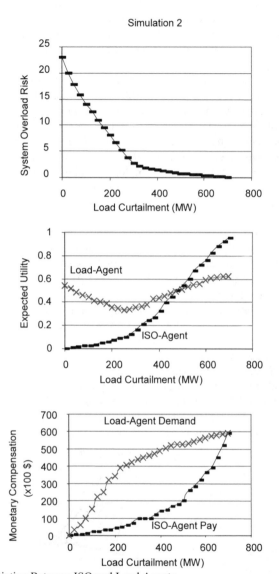

Fig. 4.6. Negotiation Between ISO and Load Agents

The multi-agent socially rational negotiation model developed in section 4.3.3 is suitable in this situation because maintenance activities would not only save money for each utility (by avoiding costly equipment failures and extend the life of electrical equipment), but also significantly improve the system reliability.

In this illustrative example, as shown in Fig. 4.7, each utility (transmission owner) employs several maintenance agents to be in charge of its electrical equipment's maintenance scheduling. The number of maintenance agents serving each utility depends on the amount of equipment the utility owns. Among the maintenance agents which belong to the same utility, they could simply rank their maintenance activities according to certain criterion, e.g. equipment failure probability, available budget etc.

While in order to carry out the maintenance activity, the corresponding maintenance agent must get approval from the ISO-Agent which is responsible for the entire system security. Using system network information, the ISO-Agent can identify allowable maintenance outage slots (number and duration) for certain time period; the requesting maintenance agents, which represent various independent organizations, negotiate among themselves to obtain more rapid approval for their maintenance activities. For example, one maintenance agent having urgent maintenance activity may be willing to negotiate with other related maintenance agents using monetary compensation in order to obtain higher priority for its maintenance activity.

We used the IEEE Three Area RTS-96 system [38] to perform a simulation with three maintenance agents (A, B, C) belonging to different organizations. They are responsible for those power transformers in each area respectively. Based on the system expected load profile, the ISO-Agent identifies time slots allowing transformer maintenance, one per time. Supposing the three maintenance agents all have pending maintenance tasks and certainly would like to have their maintenance work to be scheduled as soon as possible. So the three maintenance agents initiate negotiations among themselves. The negotiation issue is the amount of money one maintenance agent would be willing to compensate the others in exchange of scheduling its maintenance work first. For example, in the negotiation between agents A and B, A proposes a monetary compensation to B with the request of scheduling A's maintenance work ahead of B. If B does not satisfy with this offer, it will try to count-offer its own monetary compensation to A in exchange of scheduling B's maintenance work first instead.

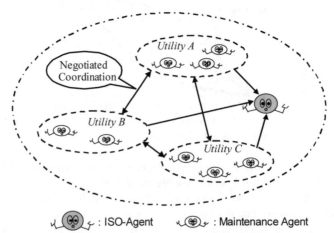

Fig. 4.7. Maintenance Scheduling based on MAS Negotiation

During the negotiation process, each agent first evaluates the expected utility of its opponent's offer, if this expected utility exceeds the utility of its own proposal, the agent will accept the offer; otherwise, it will try to counteroffer. As seen from Fig. 4.8, B accepts A's offer with the agreement of 1140 $ monetary compensation from A in exchange of scheduling A's maintenance work ahead of B's.

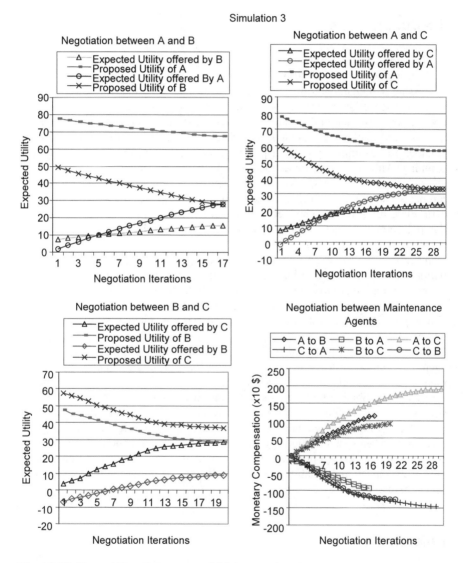

Fig. 4.8. Multi-agent Negotiation among Maintenance Agents

Similarly, B accepts C's offer with the agreement of 1260 $ monetary compensation in exchange of scheduling C's maintenance work ahead of B's; C accepts A's offer with the agreement of 1910 $ monetary compensation in exchange of scheduling A's maintenance work ahead of C's. Therefore, the overall negotiation outcome of maintenance scheduling sequence is A, C, B.

4.5 Conclusions

Multi-agent system technology is now mature enough to support negotiation as a decision paradigm among different agents. As such, it is very attractive for use in addressing a fundamental difficulty inherent to operating today's power systems where we see different stakeholders simultaneously required to compete and co-operate. We have herein illustrated two representative scenarios, one on line rating increase and another on maintenance scheduling, but there are many others, including, for example, power dispatching, emergency actions, restoration, and planning. Although agent infrastructure is available for supporting negotiated decision-making, there is significant work yet to be done in evolving negotiation protocols and corresponding analytical models. Two issues are of particular concern. First, the level of cyber-security would need to be high, and it is unclear at this point whether such security is achievable. This need could be mitigated if a multi-agent negotiation system is used only as decision support for humans rather than decision-makers themselves. Second, the degree of freedom in building negotiation models may need to be regulated, since computerized gaming in an automated negotiation environment would be highly undesirable. On the other hand, a multi-agent negotiation system offers such regulation capability, in contrast to human-based negotiated decisions in which gaming is much more difficult to detect.

Finally, we mention that we have focused our treatment of power-system multi-agent negotiation systems on implementation for real-time decision support. Yet, there is some potential that multi-agent negotiation systems hold significant promise as an investigative tool, i.e., for studying different features of distributed systems such as those found within the electric power industry. For example, it appears that multi-agent negotiation systems may play a role in examining the relation between system efficiency and the level of centralization. Here, one may conceive of comparing optimization-based simulation models that compute measures of efficiency and reliability, to similar models based on multi-agent negotiation systems.

References

[1] Wooldridge M (1999) Intelligent Agents. In: Multi-agent Systems: A Modern Approach to Distributed Artificial Intelligence, edited by Weiss G, The MIT Press, Cambridge, MA

[2]   Gulliver (1979) Disputes and Negotiations. Academic Press, New York
[3]   Chankong V, Haimes Y (1983) Multiobjective Decision Making Theory and Methodology. North-Holland.
[4]   Hobbs B, Meier P (2000) Energy decisions and the environment: A guide to the use of multicriteria methods. Kluwer, Boston, MA
[5]   Bernoulli D (1954) Expositions of a New Theory of the Measurement of Risk. English translation: *Econometrica*
[6]   Von Neumann J, Morgenstern O (1944) Theory of Games and Economic Behavior. John Wiley, New York
[7]   Keeney R, Raiffa H (1976) Decisions with Multiple Objectives: Preferences and Value Tradeoffs. John Wiley, New York
[8]   Fishburn P (1982) The Foundations of Expected Utility. D. Reidel Publishing Company
[9]   Tapan B (1997) Decision-making under uncertainty. St. Martin's Press, New York
[10] Fishburn P (1988) Nonlinear Preference and Utility Theory. The Johns Hopkins University Press
[11] Ackoff R, Sasieni M (1968) Fundamentals of Operations Research. John Wiley, New York
[12] Raiffa H (1982) The art and science of negotiation. Harvard University Press, Cambridge, MA
[13] Cross J (1969) The economics of bargaining. Basic Books Inc., New York
[14] Strauss A (1978) Negotiations: Varities, Contexts, Processes, and Social Order. Jossey-Bass Publishers, San Francisco
[15] Bartos O (1974) Process and Outcome of Negotiations. Colombia University Press, New York
[16] Rubenstein A (1982) Perfect equilibrium in a bargaining model. Econometrica 50 (1), pp 97-109
[17] Kraus S (2001) Strategic Negotiation in Multi-agent Environment. The MIT Press, Cambridge, MA
[18] Rosenschein J, Zlotkin G (1994) Rules of Encounter. The MIT Press, Cambridge, MA
[19] Huhns M, Stephens L (1999) Multi-agent Systems and Societies of Agents. In: Multi-agent Systems: A Modern Approach to Distributed Artificial Intelligence, edited by Weiss G, The MIT Press, Cambridge, MA
[20] Holsapple C, Lai H, Whinston A (1996) Implications of Negotiation Theory for Research and Development of Negotiation Support Systems. Journal of Group Decision and Negotiation, vol 6, Issue 3, pp 255-274
[21] Ehtamo H, Hamalainen R (2001) Interactive Multiple-Criteria Methods for reaching Pareto Optimal Agreements in Negotiations. Journal of Group Decision and Negotiation, vol 10, pp 475-491
[22] Kersten G (2001) Modeling Distributive and Inegrative Negotiations. Review and Revised Characterization. Journal of Group Decision and Negotiation, vol 10, pp 493-514
[23] Tajima M, Fraser N (2001) Logrolling Procedure for Multi-Issue Negotiation. Journal of Group Decision and Negotiation, vol 10, pp 217-235
[24] Weigand H, Moor A, Schoop M, Dignum F (2003) B2B Negotiation Support: The Need for a Communication Perspective. Journal of Group Decision and Negotiation, vol 12, pp 3-29

[25] Lomuscio A, Wooldridge M, Jennings N (2003) A Classification Scheme for Negotiation in Electronic Commerce. Journal of Group Decision and Negotiation, vol 12, pp 31-56

[26] Walton R, McKersie R (1965) A Behavioral Theory of Labor Negotiations. McGraw-Hill, New York

[27] Vishwanathan V, Ganugula V, McCalley J, Honavar V (2000) A multi-agent systems approach for managing dynamic information and decisions in competitive power systems. Proc. of the 2000 North American Power Symposium, Waterloo, Canada.

[28] Vishwanathan V, McCalley J, Honavar V (2001) A multi-agent system infrastructure and negotiation framework for electric power systems. Proc. of the IEEE Porto Power Tech Conference, Porto, Portugal

[29] Vishwanathan V (2001) Multi-agent system applications to electric power systems: Negotiation models for automated security-economy decisions. M.S. thesis, Dept. Elec. and Comp. Eng., Iowa State Univ., Ames

[30] http://www.recursionsw.com/products/products.asp

[31] Foundation of Intelligent Physical Agents (FIPA) specifications available at http://www.fipa.org.

[32] Barbuceanu M, Fox MS (1999) COOL: A language for describing coordination in multi-agent domains. In: First International Conference on Multi-agent systems, AAAI Press / MIT Press, pp 17-24

[33] Smith R (1980) The Contract Net Protocol: High Level Communication and Control in a Distributed Problem Solver. IEEE Trans. on Computers, vol C-29, no 12, pp 1104-1113

[34] Smith R, Davis R (1983) Negotiation as a metaphor for distributed problem solving. Artificial Intelligence 20, pp 60-109

[35] Faratin P, Sierra C, Jennings N (1997) Negotiation decision functions for autonomous agents. Int. Journal of Robotics and Autonomous Systems, vol 24, no 3-4, pp 159-182

[36] Hogg LM, Jennings NR (1997) Socially Rational Agents – Some Preliminary Thoughts. Proc 2nd Workshop on Practical Reasoning and Rationality, Manchester, pp 160-162

[37] Hogg LM, Jennings NR (1997) Socially rational agents. Proc. AAAI Fall Symposium on Socially Intelligent Agents, pp 61-63

[38] IEEE reliability test system task force of the application of probability methods subcommittee (1999) The IEEE reliability test system – 1996. IEEE Transactions on Power Systems, vol 14, no 3, pp 1010-1018

[39] Power Systems Engineering Research Center. MATPOWER, Version 2.0. available at http://www.pserc.cornell.edu/matpower/.

[40] Lutris Technologies. InstantDB, version 3.26, www.instantdb.enhydra.org.

[41] Zhang Z, Vishwanathan V, McCalley J (2002) A Multi-agent Security- Economy Decision Support Infrastructure for Deregulated Electric Power Systems. Proc. 7th International Conference on Probabilistic Methods Applied to Power Systems (PMAPS), Naples, Italy

[42] Ni M, McCalley J, Vittal V, Greene S, Ten C, Gangula V, Tayyib T (2003) Software Implementation of on-line risk-based security assessment. IEEE Transactions on Power Systems, vol 18, no 1

[43] Sparling B (2001) Transformer Monitoring: Moving Forward from Monitoring to Diagnostics. Transmission and Distribution Conference and Exposition IEEE/PES, vol 2, pp 960 -963

5 A Multi-Agent Approach to Power System Disturbance Diagnosis

Stephen D.J. McArthur, James R. McDonald, John Hossack

Institute for Energy and Environment, University of Strathclyde, Glasgow, U.K.

Protection engineers use data from a range of monitoring devices to perform post-fault disturbance diagnosis. There are a number of issues associated with this. Firstly, during storms and significant events the volume of data can overwhelm the engineers. Therefore, automated interpretation of the data, to derive meaningful information, is required. In order to achieve this, various data capture and monitoring systems must be integrated with intelligent data interpretation systems. Furthermore, extensibility must be built in to accommodate future monitoring and interpretation systems. Finally, concise and meaningful information must be provided to the end user.

This chapter discusses a Multi-Agent System (MAS) designed to provide such a flexible and scalable intelligent disturbance analysis system. It contains intelligent system based modules for alarm processing and protection analysis. Other modules provide fault record retrieval and fault record interpretation functionality. Furthermore, the chapter presents a design methodology for MAS in power engineering, considering issues such as ontology design, agent modeling, agent interaction and agent wrapping.

5.1 Introduction

The post-fault diagnosis task undertaken by protection engineers is directed at the validation of protection scheme operation. Through this, faulty components can be identified and the relevant maintenance scheduled.

Protection analysis tools have been developed by the research community to provide assistance with the tasks undertaken during post-fault diagnosis, such as alarm interpretation [1][2][3], fault record classification [4] and protection validation [5]. Although these standalone intelligent systems provide assistance with particular aspects of the diagnostic process, manual intervention is still required to collate and interpret much of the information generated. The post-fault diagnosis process can be supported through integration of these intelligent systems with improved data retrieval functions. This provides faster automated data retrieval functions, enhanced data interpretation, automated collation of related data and a common access point for information.

Multi-Agent Systems (MAS) [6] provide an effective means of integrating protection analysis tools into a flexible and scalable architecture. Previous attempts at integration [7] have resulted in inflexible architectures, where integration of new systems is restricted by the time and effort required to 'hardwire' the interfaces and establish the communications protocols. MAS overcome this by providing a standardized communications mechanism and common vocabulary between each of the integrated systems.

Recognizing the benefits of MAS, the authors have undertaken research concerning the design and implementation of a MAS, which underpins comprehensive and automated post-fault diagnosis for utility protection engineers. The multi-agent architecture resulting from this research is entitled Protection Engineering Diagnostic Agents (PEDA) [8]. This chapter will discuss protection engineering decision support requirements and the design process followed to realize PEDA.

5.2 Protection Engineering Decision Support

In order to determine the nature of the decision support required by protection engineers, the problems surrounding their current post-fault analysis process must be considered.

5.2.1 The Post-fault Analysis Process of Protection Engineers

Post-fault diagnosis commences when the protection engineer has been informed of the fault, often several hours after fault inception. The protection engineer uses a number of data sources and tools to assist with post-fault diagnosis.

SCADA (Supervisory, Control and Data Acquisition) system data is the first resource analyzed. From this, the circuits, which have experienced a fault, can be identified. Additionally, high-level indications of protection operations can be found. In order to assist with this process, previous research generated a SCADA data interpretation tool entitled the Telemetry Processor [3]. This monitors and interprets transmission SCADA. The interpretation process identifies circuit faults, or 'Incidents'. Subsequently, it groups the alarms relating to the incident and identifies the protection operation events. By examining the Telemetry Processor output the engineer can identify the circuit affected by the incident.

Having identified the affected circuit, the engineer can retrieve the fault records generated by Digital Fault Recorders (DFRs) monitoring the circuit. Data overload can result from the fact that DFRs on other circuits may trigger due to voltage dips, caused by the fault, and generate fault records of no immediate interest.

Proprietary software provided by DFR manufacturers may already have retrieved the records using autopolling, especially if several hours have elapsed since fault inception. If this is not the case, the engineer must identify the relevant DFRs and initiate retrieval using the proprietary software provided by the DFR manufacturer. However, there is no guarantee that the required records will be available, since they may have been overwritten by new fault records within the "rolling buffer".

The retrieved fault records can be used to identify the fault type, faulted phases, protection operating times and other pertinent information. Existing research had resulted in the production of a model-based protection validation and diagnosis intelligent system [9]. This uses detailed models of protection behavior to validate the protection operations recorded in the fault records against models of the protection devices. As a result, it can diagnose any protection scheme component failures.

Although the protection analysis tools used during post-fault diagnosis provide useful assistance, the provision of decision support is not optimal since the engineer must intervene to collate related information. The entire post-fault diagnosis activity is summarized in Fig. 5.1. Automation of this process would improve the analysis by ensuring timely (and prioritized) retrieval, interpretation and collation of the data and information generated by the tools.

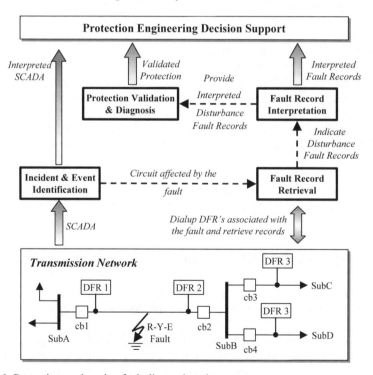

Fig. 5.1. Protection engineering fault diagnosis tasks

5.2.2 Technical Requirements for the Decision Support Functions

A number of technical requirements underlie the provision of the decision support capabilities for protection engineers. To achieve fully automated and timely post-fault diagnosis requires the integration of protection analysis tools, data sources and retrieval mechanisms. Integration alone is not sufficient as the resulting architecture should also be scalable, allowing new tools to be introduced in order to extend the decision support provision. Furthermore, the architecture must be flexible enough to deal with intermittent communications between tools. To achieve flexible and scalable integration a number of requirements must be met.

The first requirement is to provide a standardized communications protocol, which the integrated systems can use to communicate. Providing a new system conforms with the standard protocol, it can be integrated into the architecture easily. The use of an already established standard protocol avoids the time-consuming development of an architecture specific protocol.

Another communications related requirement is for a standardized communications language enabling all systems to understand exchanged messages. Previous integration attempts have focused on 'hardwiring' of system communications, with each system only understanding communications from predetermined sources and applications. This limits the architecture's scalability as the communications language must be reengineered when introducing a new system.

The final requirement is for the provision of services within the architecture, enabling new systems to discover where other applications are and the information and abilities they offer. This ensures flexibility and minimizes the need for the hardwiring of system interfaces.

Multi-agent technology provides an ideal means of achieving systems integration by wrapping up each integrated system as an 'intelligent agent'. The agent wrapper provides the integrated system with rules and knowledge enabling it to react to its environment and automate its internal functions and reasoning. Furthermore, the standardized communications mechanism and common communications vocabulary (an 'ontology' in agent terms) provided by the MAS facilitates the social interaction of the integrated systems through the agent wrappers. Finally, the provision of utility agents such as nameservers and facilitators provide the information discovery functions necessary for flexibility and scalability.

5.3 Designing the PEDA Multi-Agent System

The requirements underpinning the post-fault analysis activities, as discussed in the preceding sections, were used to drive the design of the PEDA multi-agent system.

A key issue within this research was the design of MAS for power engineering applications, and through the research activities the authors have developed the following approach and methodology to facilitate this. The key steps are:

- Requirements Capture/ Knowledge Capture
  The first step is to understand the end use of the MAS through a high level requirements specification, and to ensure that any relevant knowledge, cases, experience and rules are captured. This permits intelligent system technologies, such as rule-based reasoning and case-based reasoning to be embedded within the agent. This transforms it into an "intelligent agent".

- Task Decomposition
  The high level requirements specification and knowledge underpinning the MAS are transformed into a hierarchy of tasks and sub-tasks. It is important to note that the tasks may include functions performed by existing monitoring and analysis systems, therefore this methodology caters for legacy systems.

- Ontology Design
  The ontology is the vocabulary used by the agents to exchange information and data. As such, it is a data dictionary that supports co-operation and social ability. This is critical to the operation of the MAS.

- Agent Modeling
  From the task hierarchy and ontology design, the identification of independent and autonomous agents can be achieved. Once identified, their core functionality is designed.

- Agent Interactions Modeling
  Once the agents have been identified, the interactions between different agents must be defined. This has an impact on their core function, therefore the previous step and this one are iterative.

- Agent Behavior Functions
  The necessary software functions to allow the agent interactions are designed, and the control mechanisms designed which allow autonomy, reactiveness and pro-activeness.

- Agent Core Functionality
  Once the agent functionality has been designated, each specific agent has to be designed to perform its core function. For this application, this is a mixture of data interpretation and data retrieval functions. For other engineering applications it could be data interpretation functions, data access functions, trending functions, signal processing functions, user interface functions, etc.

5.3.1 Requirements Capture and Knowledge Capture

The design process commenced with the capturing of the knowledge necessary for the MAS to perform the post-fault diagnosis task. This required structured knowledge elicitation meetings with engineers experienced in performing post-fault diagnosis, followed by the transcribing and modeling of the elicited knowledge. The

CommonKADS [10] knowledge engineering methodology is an effective means of capturing, representing and modeling such knowledge.

To provide the knowledge and information necessary for later stages in the design process several different types of knowledge had to be identified.

Activities such as operation of proprietary fault record retrieval software and transfer of data between each fault diagnosis stage need to be identified. Both the data and software resources used during these activities were also needed, since PEDA was required to manage these resources. If post-fault diagnosis automation was to be achieved, the MAS must also mirror the reasoning processes used by the engineer to operate the software resources, interpret the output and decide which data to use in the next stage.

Having identified the activities, reasoning and resource knowledge the tasks required to perform post-fault diagnosis could be identified.

5.3.2 Task Decomposition

A MAS performs its task by decomposing the task into sub-tasks and assigning their execution to agents. Through a process of inter-agent collaboration, sub-task execution is achieved and the MAS task realized. It was therefore necessary to decompose the post-fault analysis task into its constituent sub-tasks so they could be assigned to agents.

The first stage in task decomposition was to produce a task hierarchy based solely on the transcribed knowledge. This started by decomposing the task into the sub-tasks performed by the engineer. These were identified from the knowledge transcripts as 'Identify Incidents and Events', 'Retrieve Fault Records', 'Interpret Fault Records' and 'Validate Protection Performance'. Each of these tasks was then decomposed further until the point was reached when no more detail on the task was available from the transcript.

It is important to note that the transcribed knowledge did not consider all tasks required for automation. This was particularly evident with interaction tasks. Consider the prioritized retrieval of incident related fault records. In the transcripts, the engineer identified that to enable prioritization of the 'Retrieve Fault Records' task it requires a sub-task to 'Obtain identified incidents'. However, the provision of incident information by the Telemetry Processor is not explicit since the engineer has obtained this information. It was therefore necessary to add a sub-task to the 'Identify Incidents and Events' task to 'Provide Incidents'.

The resulting task hierarchy for PEDA is presented in Fig. 5.2 with all added tasks (i.e. those not explicitly detailed within the knowledge transcripts) indicated by a *. Note that several iterations of task decomposition were required until all the tasks necessary for post-fault diagnosis automation were identified.

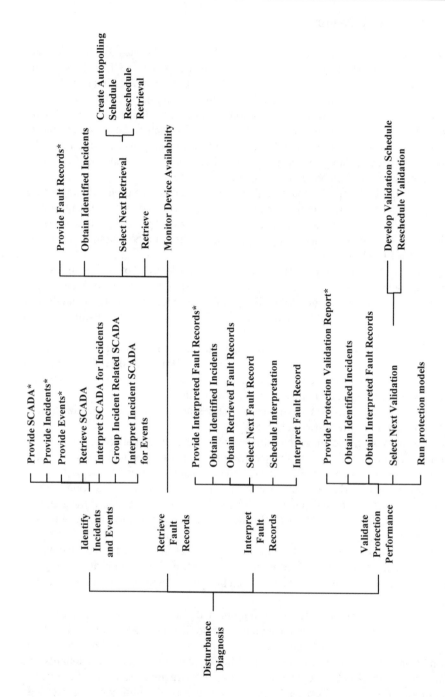

Fig. 5.2. PEDA Task Hierarchy

5.3.3 Ontology Design

The MAS architecture to be used by PEDA uses the Foundation for Intelligent Physical Agents (FIPA) Agent Communications Language (ACL) [11] to provide a communications mechanism and message structure for integration. The tasks responsible for agent interaction would use this messaging system to request the provision of a resource or provide a resource, e.g. 'Provide incidents' and 'Obtain identified incidents' shown in Fig. 5.2. In order for each agent to understand the content of these messages an architecture wide vocabulary was required called an 'ontology'. An ontology is formally defined as "explicit formal specifications of the terms in a domain and relations among them" [12].

At this stage it is important to note that a fault diagnosis ontology already existed [13], however it viewed fault diagnosis from the perspective of the control engineer and did not encompass all the terms necessary to represent post-fault diagnosis. It was therefore deemed inappropriate and design of a post-fault diagnosis ontology for PEDA commenced.

The ontology design process required the identification of the terms used by the engineer to describe the domain concepts [14]. The terms were identified by listing every term found during analysis of the knowledge transcripts and task hierarchy. Terms such as 'Incident', 'SCADA', 'Circuit' and 'Transformer' were identified.

Having listed the terms, the next stage was to identify classes describing the terms. For example, the terms 'Incident', 'Event', 'Interpreted Fault Record' and 'Protection Validation Report' can all be grouped under a class 'Information' since they are all generated by existing intelligent systems. This process was repeated until all terms were assigned to classes and the classes organized into a class hierarchy as illustrated in Fig. 5.3.

The classes enable the agents to provide and request particular types of resource. To facilitate requests for specific instances of a resource, such as a fault record from a particular DFR, the attributes of each class had to be defined. Attributes are the variables that describe each class, such as 'substation', 'date', 'time', 'device name' and 'circuit'. All sub-classes of a class inherit the attributes of that class, e.g. the attributes of 'Incident' inherit the attributes of the 'Information' class.

Following completion of the ontology the classes of information exchanged by the interaction tasks were added to the task hierarchy. This is indicated in Fig. 5.4 with resources provided by a task indicated with '⇒' and resources required by a task indicated by '⇐'.

The provision of an ontology not only provides a standardized information discovery vocabulary for use by the agents but also facilitates flexibility and scalability. For example, a new Intelligent System for providing circuit breaker trip coil condition monitoring information could be wrapped with agent functionality and integrated into PEDA. The integration would only require extension of the PEDA ontology with a new 'Data' sub-class called 'trip-coil' and a new 'Device' subclass called 'trip-coil-monitor'.

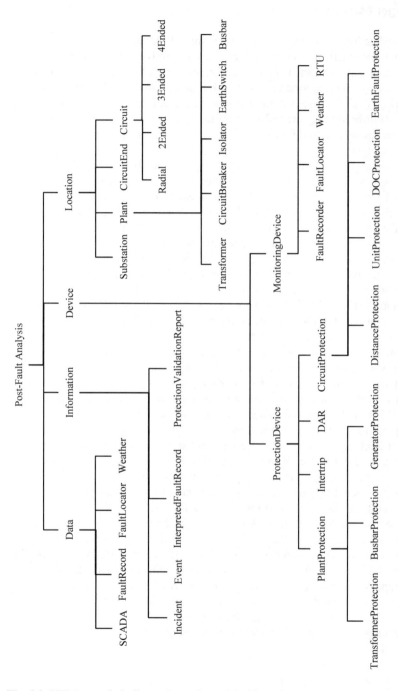

Fig. 5.3. PEDA post-fault diagnosis ontology class hierarchy

5.3.4 Agent Modeling

The identification of the required agents within a MAS and the modeling of their functional and interaction tasks is normally an iterative process where tasks are assigned to agents until each agent is performing a set of tasks which are logically related, say by resource used or type of information provided. This is not the case when using MAS for systems integration since the objective is to minimize development effort by reusing the tasks performed by existing systems. The identification of the agents required within the MAS is therefore simpler since each system becomes an agent.

To assist in agent identification the tasks performed by existing systems were grouped and the system that performs them identified from the transcripts, as illustrated for PEDA in Fig. 5.4. The agent which controls these tasks and the high-level task, or 'role', it must perform was determined by looking at the root task of the grouped tasks. For example the root task of 'Run protection models' is 'Validate Protection Performance' and this became the role of the Protection Validation and Diagnosis (PVD) agent, which employed the existing model-based protection analysis intelligent system. The other PEDA agents identified in this manner were Incident and Event Identification (IEI) and Fault Record Interpretation (FRI). The agents were also assigned all the tasks below the root task, i.e. to the right of the gray bar on the task hierarchy in Fig. 5.4.

Another consideration was whether it was possible for an agent to control the execution of existing systems. In case of the software used to retrieve fault records, it was deemed too difficult and time-consuming to obtain permission for, and undertake modifications to, the proprietary software so a new system was developed to perform these tasks. The final agent within PEDA is 'Fault Record Retrieval' (FRR) and it was assigned the task of controlling this system.

By grouping the tasks performed by existing systems the tasks still to be realized were highlighted. Each of these tasks relate to the manual tasks previously undertaken by the engineer or those required to automate post-fault diagnosis. The template in Fig. 5.5 provides a description of the role the PVD agent must undertake and functional and interaction tasks assigned to it – similar templates were also created for the IEI, FRR and FRI agents in PEDA.

Most of the functional tasks within PVD can be realized by the development of software using traditional software algorithms. However, the 'Reschedule Validation' based on received incident information' task also requires some reasoning ability in the form of rules to decide on the next validation. This decision is not only a function of the number and temporal order of received incidents but also the availability of incident related fault records. The tasks to be performed by existing intelligent systems should also be identified such as the protection analysis intelligent system that utilizes model based reasoning [9].

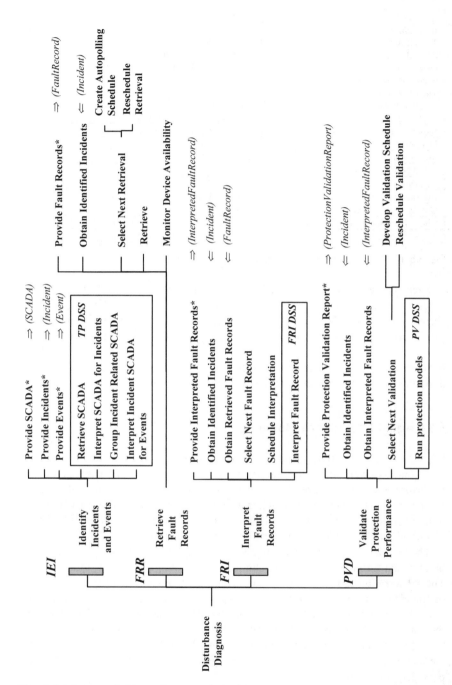

Fig. 5.4. PEDA task hierarchy extended with agent assignments, existing tools and resources exchanged during interactions

AGENT NAME:	Protection Validation and Diagnosis (PVD)	
Description:		
This agent will be responsible for the validation of protection performance and diagnosis of protection scheme component failures. This will be achieved using interpreted fault records obtained through sending query-ref and request messages to an interpreted fault record provider and the use of an existing intelligent system for protection validation. The PVD agent will also prioritise validation based on incident information received through subscription to an incident provider. When incident information is received the agent will obtain related interpreted fault records from a provider and schedule them for validation.		
Functional Tasks:	**Task Realization Method**	
Develop Validation Schedule	Algorithmic code	
Reschedule Validation	Rules, algorithmic	
Select Next Validation	Algorithmic code	
Run Diagnostic Engine	Protection validation intelligent system	
Interaction Tasks:	**Interaction Type**	**Exchanged Resource**
Obtain Identified Incidents	subscribe	Incident
Obtain Interpreted Fault Records	query-ref, request	InterpretedFaultRecord
Provide Protection Validation Reports	subscribe, query-ref	ProtectionValidationReport

Fig. 5.5. PVD Agent Modeling Template

Each interaction task uses the FIPA ACL specification to construct the messages required to exchange the resources depicted in Fig. 5.4. The specification currently contains 22 different message types of which only 'subscribe', 'confirm', 'failure', 'refuse', 'query-ref', 'request' and 'inform' are important for integration. These are described in the next section. The permissible message types for each interaction task were also specified in the template to facilitate the later design of the agent message handlers.

5.3.5 Agent Interactions Modeling

PEDA achieves automated post-fault diagnosis through agent interaction. The interaction tasks were identified and modeled, so the sequence of messages required to perform the task could be determined. This information was then used to design the agent message handlers for each task. The message sequences for each interaction are illustrated on sequence diagrams. The sequence diagram for the 'Obtain identified incidents' task of PVD is illustrated in Fig. 5.6.

The template in Fig. 5.5 had already identified that incident information was to be obtained through subscription using the 'subscribe' FIPA ACL message type. The subscription sequence has to start with the PVD agent identifying a provider of 'Incident' information by sending a 'query-ref' message with content Incident to the Facilitator. The Facilitator acts as a 'yellow-pages' and provides a list of all the abilities and resources, which agents within the MAS can provide. After a period of time, indicated by the size of the rectangle on the Facilitator's timeline, the facilitator responds with an 'inform' message containing the name of an Incident provider, namely IEI.

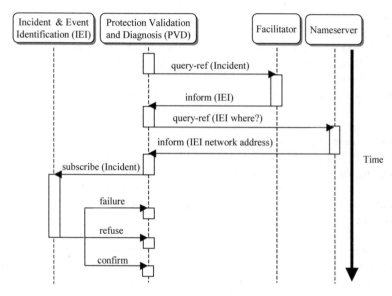

Fig. 5.6. Sequence Diagram for PVD 'Obtain identified incidents' task

The message handler then needs to identify the network location of the IEI agent so subscription communications can commence; this is achieved by interacting with the Nameserver utility agent provided by the MAS architecture. Having identified the location of IEI, the message handler then needs to send a 'subscribe' message and wait for either a 'failure', 'refuse' or 'confirm' message, which it must deal with accordingly.

The above process had to be repeated for each interaction task from both sides of the interaction e.g. 'Obtain identified incidents' and 'Provide incidents'.

5.3.6 Agent Behavior

The next stage of the multi-agent system considerations was the behavioral aspects. This was to design the 'wrappers', which provide the properties necessary for each tool to behave as an intelligent agent. The properties software must posses are: autonomy, reactivity, pro-activeness and social ability. These are the weak notions of agency [6].

Before the agent wrapper can exhibit autonomy, reactivity and pro-activeness, it must be provided with a social ability. This is achieved through the use of the FIPA ACL, ontology and message handlers.

The required message handlers were already identified during the Agent Interaction Modeling stage and are implemented as rules, which monitor the agent's incoming mailbox for messages. When a new message is received, the agent reacts by firing the rule appropriate to the received FIPA message and a function is performed, thereby providing the wrapper with reactivity. For example the message handler for handling 'inform' messages from IEI, in response to an earlier 'subscribe' message sent by the 'Obtain identified incidents' message handler of PVD, will fire when the incoming message has a FIPA ACL type 'inform'. Then it will add the 'Incident' information contained in the message to the agent's memory. When the 'Create validation schedule' task is executed all the 'Incident' information received through this mechanism will be used to create the validation schedule.

Autonomy is provided by the agent control algorithm, which controls the execution of agent tasks without user intervention. The order in which the tasks within the PEDA agents must be executed was found from the transcripts and by considering the input and output requirements of each task. The algorithm goes through a continuous loop executing each task in sequence. If a task cannot be executed due to a resource not being available the agent can request provision of the required resource from other agents in the MAS using the FIPA ACL and ontology, thereby exhibiting pro-activeness.

The method used by the wrapper to control the protection analysis tools is dependent on the type of tool. In the case of online tools such as the Telemetry Processor, little control is required since the agent only needs to start the tool and monitor for output. To add the incident and event information directly to the IEI agent's memory in the correct ontology format, minor modifications to the Telemetry Processor code were required. The modifications required to control the existing off-line tools, such as the fault record interpretation and protection validation intelligent systems, were slightly more extensive since the normal user interventions to input data and operate the tools needed to be automated.

5.4 Core Functionality of the Agents within PEDA

The previous design activities led to the specification of the overall multi-agent system and the functionality required within each agent. However, each agent must perform specific data retrieval or data interpretation functions, using existing intelligent systems or new applications. The following sections of the chapter describe how each individual agent was provided with its data interpretation or data retrieval functionality.

5.4.1 Incident and Event Identification (IEI) Agent

During periods of high network activity, such as during storms, tens of thousands of alarms can be generated over a 24 hour period. In some cases the time consuming manual interpretation of these alarms can reveal in excess of 100 separate incidents. Research conducted by the authors, concerning the automation of this process, culminated in an intelligent system employing rule-based techniques and JAVA technologies to perform automatic interpretation of transmission network

SCADA data. This identifies pertinent incidents and events [3]. The system was named the Telemetry Processor.

The inference mechanism necessary to perform incident and event identification was provided by the Java Expert System Shell (JESS) [15]. The JESS inference engine works in conjunction with core JAVA functions for managing the overall Reasoning process. The JAVA-JESS reasoning mechanism employed is illustrated in Fig. 5.7.

Three separate rule-bases exist within the Telemetry Processor responsible for incident identification, incident closure and event identification. Each rule-base contains JESS rules looking for a particular pattern of alarms and upon matching, the appropriate rules fire. Rule firings can result in the following: generation of open incident JAVA objects representing the start of an incident, closure of open incident objects when the incident has finished and generation of JAVA objects representing identified events. Within each incident and event object a concise textual summary is included for archiving and eventual viewing by an engineer.

Off-line testing revealed that 127 incidents and 2500 corresponding events were successfully identified from interpretation of 15000 alarms in 3 minutes and 17 seconds. This is a significant improvement over the number of man days required for manual interpretation.

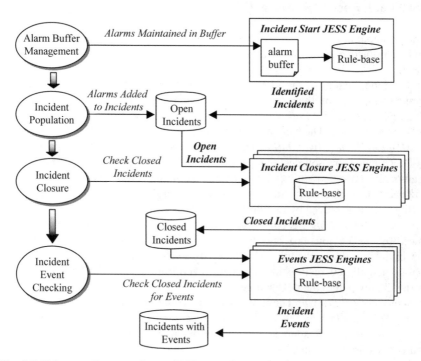

Fig. 5.7. Telemetry Processor Java – JESS reasoning mechanism

With only minor modifications to the original Telemetry Processor code, the IEI agent can start the Telemetry Processor and output the identified incidents and events in a suitable format for the IEI agent to communicate throughout the PEDA multi-agent system.

Collaboration Functionality

It is evident that the IEI agent must allow incident subscription and provide incident information to subscribed agents. This is provided by messaging rules, which monitor the IEI agents mailbox. On receiving 'subscribe' messages containing an empty Incident fact, the messaging rules automatically reply with a 'confirm' message indicating successful subscription or a 'failure' message if the subscription process failed. The IEI agent can also refuse to provide subscription services to a requesting agent by sending a 'refuse' message. This is based on permissions, to prevent inappropriate access to incident data. Successfully subscribed agents are informed of new incidents by the rules sending an 'inform' message to each subscribed agent with content being an Incident Fact containing the incident information.

5.4.2 Fault Record Retrieval (FRR)

The principal objective of this agent is to autopoll available DFRs for new records and to prioritize retrieval based on Incident information received from the IEI agent.

The functionality of the FRR agent can be described by its Java tasks and the strategy applied for their control. The implemented tasks are: *Monitor Device Availability, Create Autopolling Schedule, Select Next Retrieval, Retrieve* and *Reschedule Retrieval*. Their inputs, output and functionality are as follows:

Monitor Device Availability: This does not use any input but produces the list of DFR devices or 'sources' that are available at the time of running. At present, it reads a database containing all possible sources and selects those categorized by engineers as available. In the future a more accurate assessment of device availability will be achieved by monitoring the communications to remote DFRs.

Create Autopolling Schedule: The input is the list of available sources produced by the previous task and its output consists of a list of sources constituting the initial retrieval schedule. The order of sources dictates the retrieval order. Only when all new records have been retrieved from a particular DFR source will retrieval move on to the next source.

Select Next Retrieval: The input is the schedule produced by the *Develop Retrieval Schedule* task and its output is a request for fault record retrieval. This task selects the next DFR source to initiate retrieval from, based on the ordering of sources within the retrieval schedule.

Retrieve: The input is the fault record retrieval request issued by the previous task while its output is a reference to all the new records retrieved from the DFR source. This task establishes the connection and communications with the remote DFR and initiates retrieval of any new records. When the records have been retrieved from the DFR they are archived and the pathname of each retrieved fault record held within the FRR agent. This enables the FRR agent to reply to external requests for fault records by forwarding a reference to the records avoiding the need to pass large DFR records between agents.

Reschedule Retrieval: The input can be either an Incident fact or a FIPA 'request' message for FaultRecord facts from an external agent, in addition to the present schedule. Its output is a new schedule. The activity of this task consists of the reorganization of the list of sources to be retrieved based on a change of priorities due to external influences such as incidents or requests.

Task execution is controlled by a set of rules within the agent, which monitor for the existence or non-existence of the information required for task input. Based on this information, the appropriate rules fire and execute the required tasks.

Collaboration Functionality

The required fault record subscribe and inform functionality is implemented in the same manner as in IEI with an empty FaultRecord fact being received within a 'subscribe' message and successfully subscribed agents receiving 'inform' messages with content being a FaultRecord fact containing the fault record information.

FRR must also be able to accept 'query-ref' messages requesting particular FaultRecord facts generated by previous retrievals. FRR responds to these requests with an 'inform' message containing a set of FaultRecord facts matching the query. If no matching FaultRecord facts are available, the 'inform' message contains an empty set.

FRR must also be able to accept 'request' messages requesting retrieval of records from a particular DFR. Upon receipt of a 'request' message with the content being a FaultRecorder fact, indicating which fault records are required, the *Reschedule Retrieval* task will be called and retrieval from the requested source scheduled. When retrieval from the requested source has been completed a 'confirm' message will be sent to the requesting agent indicating completion. An 'inform' message will also be sent with content being a set of FaultRecord facts containing each retrieved fault record.

To prioritize retrieval of fault records related to an incident, the FRR agent must obtain incident information. This is achieved by querying the Facilitator and Nameserver for a provider of Incident facts. Having identified a provider, FRR initiates Incident subscription by sending a 'subscribe' message with content Incident to the provider, in this case IEI. Message handlers are used within FRR to handle the 'confirm', 'failure' or 'refuse response to subscription.

5.4.3 Fault Record Interpretation (FRI)

The principal objective of this agent is to obtain and interpret fault records relating to an Incident to generate information of interest to the protection engineers.

Tasks are implemented in a similar manner to the FRR agent with rules controlling task execution. The implemented tasks are: *Schedule Interpretation, Select Next Fault Record* and *Interpret Fault Record*.

Schedule Interpretation: The inputs to this task are the sets of retrieved fault record information provided by the FRR agent following a retrieval request and the incident information provided by IEI. The FRI agent is solely reliant on information received from other agents and does not have an independent task to perform, like autopolling in the FRR. This task outputs an interpretation schedule.

Select Next Fault Record: The input is the interpretation schedule and the output is a request for fault record interpretation. The decision on which record to interpret is based on the ordering in the interpretation schedule.

Interpret Fault Record: The input to this task is the fault record interpretation request issued by the previous task, which specifies the pathname of the fault record to interpret. This task is the principal task within the FRI agent and is responsible for interpreting the fault record and extracting information of relevance to protection engineers.

The types of information extracted are:

- Fault inception and clearance time; and
- Protection scheme device operating times.

The fault inception time is identified by determining the earliest time of a phase current rising significantly above normal. Fault clearance time is determined by identifying the latest time when zero current is detected on all phases. Protection scheme device operating times are determined by interpreting the digital channels of the fault record and extracting the time periods for which each monitored digital is in a given state.

Future work will see extension of the interpretation functionality to provide additional information such as fault type and circuit breaker duty.

Collaboration Functionality

The interpreted fault record subscribe and inform functionality is implemented in the same manner as in FRR with InterpretedFaultRecord facts being passed within the 'subscribe' and 'inform' messages.

To prioritize interpretation of fault records related to an incident the FRI agent must subscribe to an Incident provider in the same manner as in FRR. To obtain the fault records relating to an Incident the FRI agent identifies a provider of FaultRecord facts by querying the Facilitator and Nameserver. Having identified a provider the FRI agent sends a 'query-ref' and, if required, a 'request' message to the provider, in this case FRR.

The ability to accept 'query-ref' and 'request' messages is also provided in FRI using a similar mechanism to FRR. To process 'request' messages the FRI agent will need to obtain FaultRecord facts which can in turn be interpreted and the requested InterpretedFaultRecord facts generated.

5.4.4 Protection Validation and Diagnosis (PVD)

The principal objective of this agent is to validate the operation of a protection scheme during an incident and to diagnose any failures in the protection scheme components.

Tasks are implemented in a similar manner to the FRR and FRI agents with rules controlling task execution. The implemented tasks are: *Develop Validation Schedule*, *Reschedule Validation*, *Select Next Validation* and *Run Diagnostic Engine*.

Develop Validation Schedule: The inputs to this task are any received inform messages containing Incident facts. If no incidents are currently being validated a new schedule is created and the received Incident added. Alternatively, the received fact is added to the existing schedule. This task is also responsible for identifying the DFRs on the incident circuit and obtaining the associated Interpreted-FaultRecord facts required for validation. To obtain the required facts a fault recorder database is queried for devices on the incident circuit and the identified FaultRecorder facts output for the 'Obtain Interpreted Fault Records' interaction task to pick up and process. The task also outputs the updated validation schedule.

Reschedule Validation: The input to this task is each received 'inform' message containing InterpretedFaultRecord facts resulting from the 'Obtain Interpreted Fault Records' interaction task. Each received fact is recorded against the related incident and only when a response has been received related to each device on the circuit is the incident marked as being ready for validation. At this stage this incident is moved to the top of the validation schedule. The output of this task is the updated validation schedule.

Select Next Validation: The input to this task is the validation schedule. The first incident marked as ready for validation is selected and the associated Incident and InterpretedFaultRecord facts output.

Run Diagnostic engine: The input to this task is the Incident fact to be validated and each related InterpretedFaultRecord. The InterpretedFaultRecord fact contains the FaultRecord fact, which means that the PVD agent has access to the raw data from the fault record for its diagnostic process. This task is the principal task within the PVD agent and is responsible for validating the operation of the protection scheme on the incident circuit. Following validation a protection validation report is generated providing a validation of protection scheme component operations and a diagnosis of any failures. This report is output as a Protection-ValidationReport fact for other agents to obtain if required.

Collaboration Functionality

To provide protection validation reports the subscribe and inform functionality is implemented in the same manner as in FRI with ProtectionValidationReport facts being passed within the 'subscribe' and 'inform' messages. The ability to accept 'query-ref' messages for protection validation reports is also provided.

To prioritize incident protection performance PVD follows the same Incident subscription approach as adopted in FRR and FRI. When PVD is informed of an Incident the 'Obtain Interpreted Fault Records' interaction task is executed. This task sends the FRI agent 'query-ref' and 'request' messages for InterpretedFault-Record facts in the same manner as adopted by FRI for FaultRecord retrieval.

5.5 Case Study of PEDA Operation

To illustrate the PEDA approach to disturbance diagnosis an actual power system disturbance will be used as a case study. Note that, to aid clarity, the network diagram in Fig. 5.8 has been simplified and the data presented in Fig. 5.9 reduced. A diagram of PEDA, illustrating the agents, is presented in Fig. 5.10 with the inter-agent collaborations shown in Table 5.1 for reference.

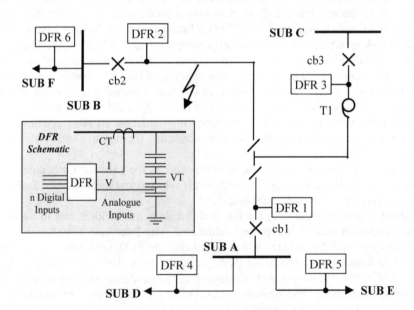

Fig. 5.8. Case Study – Transient Fault on circuit SUBA / SUBB

SCADA Alarms and Indications

13:54:14:720 SUBA SUBB	Unit Protection Optd	ON
13:54:14:730 SUBA SUBB	Trip Relays Optd	ON
13:54:14:730 SUBF	Battery Volts Low	ON
13:54:14:750 SUBF	Battery Volts Low	OFF
13:54:14:760 SUBA SUBB	Second Intertrip Optd	ON
13:54:14:760 SUBA cb1	OPEN	
13:54:14:790 SUBB SUBA	Distance Protection Optd	ON
13:54:14:800 SUBC T1	Trip Relays Optd	ON
13:54:14:800 SUBD	Pilot Faulty	ON
13:54:14:800 SUBB SUBA	Trip Relays Optd	ON
13:54:14:810 SUBD	Pilot Faulty	OFF
13:54:14:810 SUBB SUBA	Unit Protection Optd	ON
13:54:14:810 SUBE	Metering Alarms	ON
13:54:14:820 SUBB cb2	OPEN	
13:54:14:830 SUBA SUBB	Autoswitching in Progress	
13:54:14:850 SUBC cb3	OPEN	
13:54:14:860 SUBB SUBA	Second Intertrip Optd	ON
13:54:14:860 SUBB SUBA	First Intertrip Optd	ON
13:54:14:890 SUBB SUBA	Autoswitching in Progress	
13:54:14:900 SUBC T1	Autoswitching in Progress	

13:54:35:470 SUBC T3	Autoswitching Complete	
13:54:35:490 SUBC cb3	CLOSED	
13:54:37:710 SUBB cb2	CLOSED	
13:54:38:210 SUBA SUBB	Autoswitching Complete	
13:54:38:300 SUBA cb1	CLOSED	

DFR Records

13:54:14.740 DFR1 Triggered on Protection
13:54:14.740 DFR4 Triggered on low volts
13:54:14.740 DFR5 Triggered on low volts
13:54:14.810 DFR2 Triggered on Protection
13:54:14.810 DFR6 Triggered on low volts
13:54:14.850 DFR3 Triggered on CB OPEN

Fig. 5.9. Case Study Alarm and DFR data

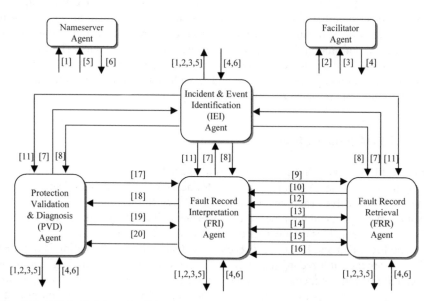

Fig. 5.10. PEDA Collaboration Diagram

Table 5.1. Interactions within the PEDA multi-agent system

ID	PEDA Collaborations
1	Register agents' network location
2	Register types of information agent can provide and its abilities
3	'query-ref' for a provider of a particular information type or ability
4	'inform' of the provider's name
5	'query-ref' for the network location of a given agent
6	'inform' of the network location of a given agent
7	'subscribe' to IEI for Incident information
8	'confirm', 'refuse' or 'failure' response to Incident subscription
9	'subscribe' to FRR for FaultRecord information
10	'confirm'. 'refuse' or 'failure' response to FaultRecord subscription
11	'inform' of new Incident information
12	'inform' of new FaultRecord information due to subscription
13	'query-ref' for a particular FaultRecord
14	'inform' of FaultRecords in response to 'query-ref'
15	'request' retrieval of fault records from a FaultRecorder
16	'confirm' that retrieval request is complete
17	'query-ref' for a particular InterpretedFaultRecord
18	'inform' of InterpretedFaultRecords matching 'query-ref' content
19	'request' interpretation of fault records from a particular FaultRecorder
20	'confirm' that interpretation request is complete.

Manual retrieval and interpretation of the SCADA and DFR data by a protection engineer revealed that the disturbance was caused by a transient fault on the SUB A / SUB B / SUB C circuit. The protection scheme detected the fault, isolated and restored the circuit using autoswitching. Ignoring that the engineer was not made aware of the disturbance until arriving at work the next morning, the total retrieval and interpretation time was in the order of 2-3 hours. The following discussion will illustrate how PEDA can reduce the data retrieval and interpretation overhead on engineers by novel and flexible automation of the process using MAS to integrate retrieval and interpretation systems.

5.5.1 PEDA Initialisation

Before PEDA can commence with disturbance diagnosis it must go through an initialization phase. Assuming that the Nameserver and Facilitator agents have started this involves each agent within PEDA registering their location with the Nameserver agent and the type of resources and abilities they can provide with the Facilitator agent. This is shown as collaborations 1 and 2 in Table 5.1.

The FRR, FRI and PVD agents each know that incident information is required for prioritizing fault record retrieval, fault record interpretation and validation of incident protection scheme operations. They each obtain this information by sending a 'query-ref' message to the Facilitator asking for an agent who can provide

Incident information. The Facilitator responds to each 'query-ref' with an 'inform' message containing the IEI agent name. This is shown as collaborations 3 and 4 in Table 5.1.

Before subscription to the IEI agent can proceed the FRR, FRI and PVD agents must determine its location. This is achieved by sending a 'query-ref' message with content IEI to the Nameserver. The Nameserver responds to each 'query-ref' with an 'inform' message containing the network location of the IEI agent. This is shown as collaborations 5 and 6 in Table 5.1.

Having identified the IEI agent as a provider of Incident information and its network location the FRR, FRI and PVD agents each send a 'subscribe' message to it requesting Incident subscription. It responds to each 'query-ref' with either a 'confirm', 'refuse' or 'failure' message depending on the success of the subscription and whether it's permissible for the requesting agent to subscribe for Incident information. For the remainder of the case study it will be assumed that each 'subscribe' message results in successful subscription. This subscription process is shown as collaborations 7 and 8 in Table 5.1.

To subscribe for FaultRecord information the FRI agent employs a similar procedure of identifying and locating a FaultRecord provider before sending a 'subscribe' message. This is shown in collaborations 9 and 10 in Table 5.1.

5.5.2 Disturbance Diagnosis Process

Before discussing how PEDA performs disturbance diagnosis it is important to note that the PEDA agents perform their functions concurrently. So whilst the IEI agent is retrieving and interpreting new SCADA data, the FRR agent is autopolling DFRs and retrieving any new fault records, the FRI agent is interpreting any new fault records retrieved by FRR and the PVD agent is conducting protection validation of any previously identified incidents. Only when an agent's internal reasoning determines that inter-agent communications is necessary are messages sent between agents.

On entering PEDA, the IEI agent will connect to the real-time SCADA alarms database and start the Telemetry Processor interpreting the alarms. Given the alarm stream in Fig. 5.9, the incident identification rules will recognize the 'Unit Protection Optd', 'Trip Relays Optd' and 'cb1 OPEN' alarms on the SUBA SUBB circuit. These indicate the start of an incident and generate an incident object. Any alarms on this circuit, which are subsequently received, will be added to the open incident object, thereby grouping incident alarms. Incident closure rules will then interpret the incident alarms for incident closure, recognizing a complete autoswitching sequence on the circuit. This will close the incident and generate the incident summary indicated at the top of Fig. 5.11. Event identification rules will then further interpret the incident alarms and generate information on events of interest to the engineers. An example of four of the generated events is presented in Fig. 5.11 below the incident summary.

Incident:	**"13:54:14.720 SUB A / SUB B / SUB C Autoswitching Sequence Complete"**
Events:	13:54:14.720 "Unit Protection Operated Successfully at SUBA -> SUBB
	13:54:14.810 "Unit & Distance Protection Operated Successfully ay SUBB -> SUBA
	13:54:14.860 "1st and 2nd Intertrips received at SUBB from SUBA
	13: 54:38.300 "Distance Protection at SUBA -> SUBB failed to operate

Fig. 5.11. Sample of Telemetry Processor output from within IEI

Having identified an incident, the IEI agent's message handling rules will automatically inform all subscribed agents of the new incident. In this case the FRR, FRI and PVD agents will each receive an 'inform' message indicating the date, time, circuit and a summary of the incident. This is collaboration 11 in Table 5.1.

Having subscribed to the IEI agent, the FRR *Reschedule Retrieval* task is executed, prioritizing retrieval of records from DFRs on the same circuit as the incident, i.e. DFR1, DFR2 and DFR3. The retrieval process may take anywhere from tens of seconds to several minutes depending on the quality of the communications to the remote fault recorders.

The FRI agent will also have received the Incident information and will send a 'query-ref' message to the FRR agent for the FaultRecords related to the Incident – collaboration 13 in Table 5.1. Since retrieval is still underway and no FaultRecords are yet available an 'inform' message with an empty set is sent back to the FRI agent – collaboration 14 in Table 5.1. Upon receipt of this, the FRI agent knows that it must obtain the Incident FaultRecords so sends a 'request' for retrieval to the FRR agent for each FaultRecorder on the incident circuit – collaboration 15 in Table 5.1. When the FRR agent has completed retrieval from the requested devices a 'confirm' message is sent to the FRI agent informing it of completed retrieval - collaboration 16 in Table 5.1.

To obtain the FaultRecords retrieved from DFR1, DFR2 and DFR3, the FRI agent sends a 'query-ref' message to the FRR agent. This agent responds with an 'inform' message containing each corresponding FaultRecord fact. These are collaborations 13 and 14 in Table 5.1. Having obtained the FaultRecords FRI can schedule each one for interpretation.

The PVD agent also received the same Incident inform message as the FRR and FRI agents. Therefore, it proceeds to obtain the InterpretedFaultRecord facts relating to the Incident as follows. The 'Obtain Interpreted Fault Records' task identifies that InterpretedFaultRecord facts will be required for the fault records generated by fault recorders DFR1, DFR2 and DFR3. It sends a 'query-ref' to the FRI agent. At this point in time the FRI agent will have requested, but not received, the fault records from the FRR agent. Consequently, it sends an 'inform' message with an empty InterpretedFaultRecord fact to the PVD agent – collaborations 17 and 18 in Table 5.1. The PVD agent will then send a 'request' message to the FRI agent requesting interpretation of fault records from FaultRecorders DFR1, DFR2 and DFR3. When the FRI agent has received and interpreted the fault records a 'confirm' message will be sent to the PVD agent in response to each 'request' message. These are collaborations 19 and 20 in Table 5.1.

To obtain the requested InterpretedFaultRecords, the PVD agent sends a 'query-ref' message to the FRI agent. The FRI agent responds with an 'inform' message containing each corresponding InterpretedFaultRecord fact. These are collaborations 17 and 18 in Table 5.1. Having obtained the FaultRecords, the FRI agent can schedule each for interpretation.

At the completion of the interactions between the agents, full interpretation of the SCADA system data and fault records will be achieved. This will indicate the incidents and events that had taken place, and will provide an assessment of the performance of the protection schemes, including diagnosis of device failures and slow operations. A single user interface will be provided to allow the protection engineers to view the conclusions of the PEDA multi-agent system.

5.6 Conclusions

The opportunity to optimize the provision of post-fault diagnosis decision support assistance to protection engineers through the integration of protection analysis tools has been recognized. Multi-agent technologies have been identified as an effective means of achieving this integration within a flexible and scalable architecture.

The core of this chapter has outlined how the authors have integrated the protection analysis tools used during post-fault diagnosis into a MAS entitled Protection Engineering Diagnostic Agents (PEDA). Tools for SCADA interpretation, fault record retrieval, fault record interpretation and protection validation have all been provided with agent functionality through the agent wrappers. Integration has been achieved through the use of the FIPA ACL, utility agents provided by the MAS and the post-fault diagnosis ontology designed by the authors.

Integration using multi-agent technologies has provided a flexible and scalable PEDA architecture open to the introduction of new agents. As new tools become available the same design process can be used to provide the tools with the autonomy, pro-activeness, reactivity and social properties necessary to function as intelligent agents within PEDA. With minor extensions to the post-fault diagnosis ontology, these new agents can enter PEDA and further enhance the decision support assistance provided.

References

[1] Vale ZA, Ferandes MF (1998) Better KBS for real-time applications in power system control centers: the experience of the SPARSE project. Computers in Industry, vol 37, pp 97-111

[2] Jung J, Liu CC, Hong M, Gallanti M, Tornielli G (2002) Multiple Hypotheses and Their Credibility in On-Line Fault Diagnosis. IEEE Transactions on Power Delivery, vol 16, no 2, pp 225-230

[3] Hossack JH, Burt GM, McDonald JR, Cumming T, Stoke J (2001) Progressive Power System Data Interpretation and Information Dissemination. Proc. 2001 IEEE Transmission and Distribution Conference and Exposition

[4] Kezunovic M, Rikalo I, Vesovic B, Goiffon SL (1998) The Next Generation System for Automated DFR File Classification. Fault & Disturbance Analysis Conference,

[5] Bell SC, McArthur SDJ, McDonald JR, Burt GM, ea. (1998) Model Based Analysis of Protection System Performance. IEE Proc. Gen. Trans. & Dist., vol 145, no 3, pp 547-552

[6] Wooldridge M, et al (1995) Intelligent Agents: Theory and Practice. The Knowledge Engineering Review, vol 10, no 2, pp 115-152

[7] McArthur SDJ, Bell SC, McDonald JR, ea (1998) The Development of an Advanced Suite of Data Interpretation Facilities for the Analysis of Power System Disturbances. CIGRE 1998, Paris, France

[8] Hossack JA, Menal J, McArthur SDJ, McDonald JR (2003) A Multi-Agent Architecture for Protection Engineering Diagnostic Assistance. IEEE Transactions on Power Systems, (accepted and awaiting publication)

[9] Davidson E, McArthur SDJ, McDonald JR (2003) A Tool-Set for Applying Model Based Reasoning Techniques to Diagnostics of Power Systems Protection. IEEE Transactions on Power Systems, (accepted and awaiting publication)

[10] Schreiber G, Akkermans H, Anjwierden (1999) Knowledge Engineering and Management: The CommonKADS Methodology. MIT Press

[11] FIPA Communicative Act Library Specification. XC00037H, http://www.fipa.org/repository/index.html

[12] Uschold M, Gruninger M (1996) Ontologies: Principles, Methods and Applications. Knowledge Engineering Review, vol 11, no 2

[13] Bernaras A, et al (1996) An Ontology for Fault Diagnosis in Electrical Networks. ISAP 1996, pp 199-203

[14] Noy NF, et al (2001) Ontology Development 101: A Guide to Creating Your First Ontology. KSL Technical Report, KSL-01-05

[15] Friedman-Hill EJ (2003) Jess, The Java Expert System Shell. http://heerzberg.ca.sandia.gov/jess, version 6.1a5

6 A Multi-agent Approach to Power System Restoration

Takeshi Nagata [1], Hiroshi Sasaki [2]

[1] Dept. of Information and Intellectual Systems Engineering, Hiroshima Institute of Technology
[2] Dept. of Electrical Engineering, Hiroshima University

This chapter proposes a multi-agent approach to power system restoration. The proposed system consists of a number of *bus agents (BAGs)* and a single *facilitator agent (FAG)*. The *BAG* is developed to decide a suboptimal target configuration after a fault occurrence by interacting with other *BAGs* based on only locally available information, while the *FAG* is to act as a manager in the decision process. The interaction of several simple agents leads to a dynamic system, allowing efficient approximation of a solution. Simulation results have demonstrated that this method is able to reach sub-optimal target configurations, which are favorably compared with those obtained by a mathematical programming approach.

6.1 Power System Restoration Methods

When electric power supply is interrupted by a fault, it is imperative to restore the power system promptly to an optimal target configuration after the fault. The problem of obtaining a target system is referred to as power system restoration. To obtain the target configuration, various approaches have been proposed, which can be roughly classified into four categories: heuristics [1-5], expert systems [6-11], mathematical programming [12], and soft computing [13].

Heuristics and expert systems have been used extensively in industries, but they both have their own deficiency with respect to the optimality of solutions. Mathematical programming, on the other hand, is able to obtain an optimal solution, but it needs some engineering judgment in formulating restoration problems due to its complexity. Also, its long execution time may sometimes make feel mathematical programming in practical considering the time constraints on site. Although soft computing methods are easy to implement, they cannot obtain the optimal solutions in the true sense. Also, they need long computation time to find a solution.

Currently, agents are the focus of intense attention in many fields in computer science and artificial intelligence. In fact, agents are being used in a wide variety of applications. Many important computer applications such as planning, process control, communication network configurations, and concurrent systems will benefit from a multi-agent system approach [14-18]. A multi-agent system is a computational system in which several agents cooperate to achieve some tasks. Multi-agent models are oriented towards interactions, collaborative phenomena, and autonomy.

This chapter proposes a multi-agent approach to restoration of a local network, which consists of a number of *Bus Agents* (*BAGs*) and a single *Facilitator Agent* (*FAG*). The *BAG* is developed to decide a sub-optimal target configuration after a fault occurrence by interacting with other *BAGs*, while the *FAG* is to act as a manager in the decision process.

The multi-agent system has the following characteristics: In order to realize efficient processing, the types of agents created are restricted to only two types of agents, *BAG* and *FAG*, in the proposed multi-agent system. The *FAG* is developed by making use of the singleton design pattern [19], which ensures that only a single instance of the *FAG* exists at the same time. It is postulated that a *BAG* communicates with only its neighboring *BAGs* and furthermore the number of times of communications is limited for efficiency reasons. The *BAG* has the following simple negotiation strategies:

- If there are a plural number of branches that can energize a certain bus, the *BAG* allocated to this particular bus selects a branch with the largest amount of available restoration power.
- If the amount of available power for restoration is insufficient, the *BAG* tries to restore the bus in charge of by negotiating with its neighboring *BAGs*.
- If a load must be shed because of insufficient power, the *BAG* cuts off the load connected to its own bus as small as possible.

An object-protocol like KQML (Knowledge Query and Manipulation Language [20]) is used as a communication message scheme between the agents.

In order to verify the performance of the proposed method, many simulations are carried out using a model system that consists of 8 substations and 14 buses. Though this particular example system is small if applied to power system analysis algorithms such as load flow, restoration problems are belonging to a class of combinatorial optimization and therefore even an example system of this size is quite practical.

Simulation results obtained have demonstrated that the proposed multi-agent system is able to find a sub-optimal target configuration in all cases. It is noted that the proposed approach could reach the right solutions by using only local information. The results obtained in this chapter may suggest the practicability of the multi-agent approach for handling a much larger system.

6.2 Power System Restoration Model

The objective of the mathematical model of power system restoration is to maximize the capacity of the served loads:

$$max \sum_{k \in R} L_k \cdot y_k \tag{6.1}$$

where L_k is the load at bus k, y_k is the decision variable of expressing its status (y_k = 1: restored; $y_k = 0$: not restored), and R denotes the set of de-energized loads.

Typical constraints associated with the restoration model are taken into account in this study:

(a) Limit on the capacity of available power source for restoration:

$$\sum_{e \in F_q} P_e \cdot x_e \leq G_q \qquad (q \in S) \tag{6.2}$$

where P_e is the power flow on the directed branch e (we assume $P_e \geq 0$), x_e the decision variable of branch e ($x_e = 1$: e is included in the restoration path; $x_e = 0$: otherwise), F_q the set of branches with starting node q, G_q the restoration power from the energized bus q, and S the set of energized buses that can be connected to de-energized area.

(b) Power balance between supply and demand:

$$\sum_{k \in T_i} P_k - \sum_{k \in F_i} P_k - L_i \cdot y_i = 0 \qquad (i \in N) \tag{6.3}$$

where T_i is the set of branches incident to bus i, F_i the set of branches with originating from bus i, L_i the load at bus i, and N the set of buses.

(c) Limits on branch power flow:

$$|P_k| - U_k \leq 0 \quad (k \in B) \tag{6.4}$$

where P_k denotes the power flow of branch k, U_k the capacity of branch k, and B the set of directed branches.

(d) Constraint on radial configuration:
This constraint means that an obtained target configuration must be radial, and is used mandatory in the actual power system operations. To insure a radial configuration, the total number of branches incident to bus i must be at most unity.

$$\sum_{k \in T_i} x_k \leq 1 \qquad (i \in N) \tag{6.5}$$

6.3 Multi-agent Power System Restoration Architecture

In this section, the multi-agent architecture for power system restoration using the object-oriented design technique is introduced. Since it is essential to develop an efficient solution, we restrict the types of agents and the number of communications between agents.

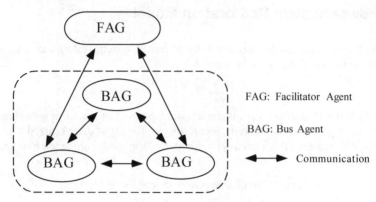

Fig. 6.1. Architecture of the proposed multi-agent system

The proposed multi-agent restoration system, as shown in Fig. 6.1, consists of a single *FAG* and a number of *BAGs*. The number of *BAGs* is the same as the number of buses. In other words, one *BAG* is allocated to each bus.

6.3.1 Bus Agents

The purpose of the bus agents (*BAG*) is to restore the load directly connected to its associated bus. The *BAG* has a set of simple rules for restoring its load:

- If there are several available points for restoration or boundary points between the energized and de-energized area, then the *BAG* has to restore its own bus through one of these points with the maximum capacity.
- If the *BAG* succeeds the restoration, it tries to make negotiations with the neighboring *BAGs*.
- If the power available for restoration is insufficient to restore all the load of the *BAG*, then it starts negotiations with the neighboring *BAGs* trying to energize the unserved load as much as possible.
- If the load shedding is unavoidable, the *BAG* has to cut off the load as little as possible.

When a fault occurs in a power system network, the *BAG* corresponding to a de-energized bus asks the *FAG* to restore its own bus. Since many buses are de-energized because of a fault, many messages are sent to the *FAG*. The first come message is the trigger to the proposed multi-agent system, that is, the initiation of its function. Responding to the first received message, the *FAG* sends out a start message to the *BAG*. On receiving the message, it starts by itself the communication with the neighboring *BAGs*.

6.3.2 Facilitator Agent

The facilitator agent (*FAG*) is a special purpose agent that facilitates the negotiation process of the multi-agent system. First, the *FAG* classifies the de-energized buses into several groups, each of which has the same voltage level. Then, the *FAG* selects one of the *BAGs* in the highest voltage group, and sends out a start message. The *FAG* repeats the same task in parallel to other de-energized groups. This means the parallel processing is carried out to restore de-energized networks. In order to ensure that only a single instance of the *FAG* exists at any time, the singleton design pattern [18] is used.

6.3.3 Negotiation Process between Bus Agents

The negotiation process is one of the key processes for the multi-agent system to successfully attain its goal. In the following, it is described what are negotiations in the restoration process and how a *BAG* makes negotiations with other *BAGs*. For the ease of understanding of the negotiation process, we explain the restoration process using a model network depicted in Fig. 6.2, on which we postulate a fault to illustrate the performance of the proposed multi-agent system. Under this particular fault, the hatched area shown in Fig. 6.2 has lost power. The line between B s/s and D s/s is tripped off because of the assumed fault and three loads, R={L2, L5, L6}, are to be re-supplied by the agent system. In this particular case, BUS 3 and BUS 9 have power available for restoration, which is denoted as S={BUS 3, BUS 9}.

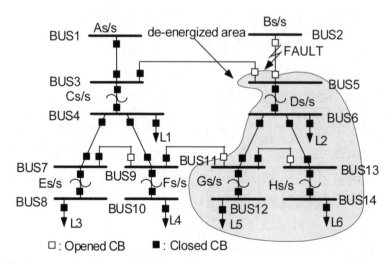

Fig. 6.2. An example of the post fault network

A sequence of the negotiations is shown in Fig. 6.3. In this figure, the hatched rectangular boxes represent *BAGs* corresponding to de-energized buses, while the other rectangular boxes are *BAGs* associated with energized buses. The fault has made the following buses de-energized: BUS 5, BUS 6, BUS 11, BUS 12, BUS 13, and BUS 14. De-energized bus BUS i corresponds to Bus Agents *BAGi* as shown in Fig. 6.3 (1). In this figure, numerals in the parentheses adjacent to *BAG3* and *BAG9* represent the available power, and figures set beside rectangular boxes signify the load of *BAGs*. *BAG12* starts a negotiation task with *BAG11* because it can only be energized through *BAG11*. The relation between *BAG13* and *BAG14* is the same as the foregoing. Therefore, *BAG5*, *BAG6*, *BAG11*, and *BAG13* respectively send a message to the *FAG* to ask to restore their own buses.

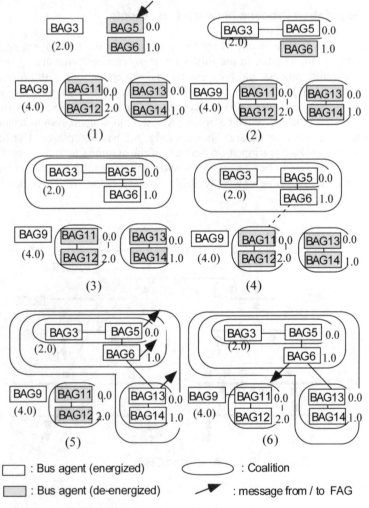

Fig. 6.3. An example of the negotiation process

The *FAG* saves these requirements to the de-energized bus list, DEBList, out of which the *FAG* selects a certain starting *BAG* and sends it a message to start communications for restoration. In this case, *BAG5* is selected since it has the highest voltage level. If there are plural available buses at the highest voltage level, the *FAG* selects one with the largest available power.

Then, *BAG5* makes negotiations with *BAG3* (Fig. 6.3 (2)), since the available power (2.0) is greater than the load at *BAG3* (0.0). Next, *BAG5* negotiates with *BAG6* (Fig. 6.3 (3)). *BUS 6* has two neighboring *BAGs*, that is, *BAG11* and *BAG13*. It is assumed that *BAG6* tries to make negotiations with *BAG11* first, but *BAG11* rejects the request (Fig. 6.3 (4)) since the current available power (1.0) is insufficient to supply its downstream load (2.0). *BAG6*, therefore, tries to negotiate with *BAG13*, and this time the request is accepted because of the sufficient available power. As *BAG13* has no neighboring *BAGs* except *BAG14* at further end, *BAG13* gives *FAG* notice that all the negotiation process has been terminated. *BAG6* and *BAG5* also send the termination message to *FAG* (Fig. 6.3 (5)).

When the *FAG* receives the termination messages from *BAG13*, *BAG6*, and *BAG5*, it removes these *BAGs* from the *DEBList*. BUG11 is still remaining in the *DEBList*. Then the *FAG* sends a message to *BAG11* to start the negotiation process in the same way as *BAG5* undertook (Fig. 6.3 (6)). After these negotiation processes have been completed, a target configuration for restoration is obtained as shown in Fig. 6.4. It requires two closing and one opening switch operations in order to restore all the loads.

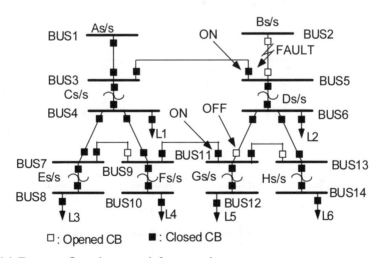

Fig. 6.4. Target configuration network for restoration

6.4 Simulation Results

6.4.1 Simulation Conditions

In order to demonstrate the effectiveness of the proposed approach, it has been applied to a model network, which consists of 8 substations (A s/s - H s/s) and 14 buses as shown in Fig. 6.5. Loads are indicated by arrows together with their magnitudes. A pair of figures on a branch shows its line flow capacity (upper row) and the amount of power flow (lower row). A square illustrates the status of a circuit breaker {black square (■): closed; white square (□): open}. It is assumed that a fault has occurred at the point shown by ✕, and the half-toned buses and loads are de-energized.

This restoration problem is difficult to solve because of a large variety of possible restoration strategies. The computer program was written in JAVA using JDK 1.1. The computer used for simulations was a PC (550 MHz).

6.4.2 Simulation Results

A large number of simulations were carried out on this model network with changing conditions.

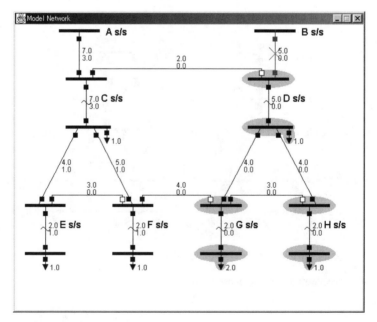

Fig. 6.5. Model network

Full restoration (Case 1)

This is a case in which all loads can be restored, since the amount of available power becomes 6.0 (2.0 from *BAG3*; 4.0 from *BAG9*), while the sum of de-energized loads is 4.0. Fig. 6.6 shows the simulation results: the optimal target configuration network and the list of switching operation sequence. The negotiation process has turned out to be the same as described in sub-section 6.3. The first two steps in the list correspond to the tripping of the faulted line, and hence the remaining three switching operations (open: 1, close: 2) are required to restore all the loads.

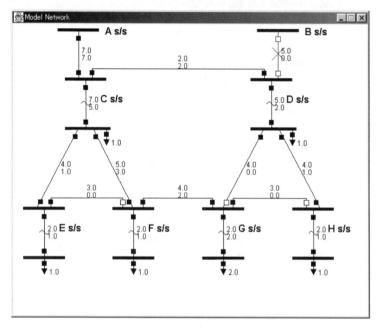

(a) Target network

NO	Station	Operation Contents	
1	Bs/s	B-D Line	Open
2	Ds/s	B-D Line	Open
3	Gs/s	D-G Line	Open
4	Ds/s	C-D Line	Close
5	Gs/s	F-G Line	Close

(b) Switching sequence

Fig. 6.6. Case 1: Full restoration

Partial restoration of loads (Case 2)

In the next, we discuss a case of partial restoration where the amount of available power falls short of the sum of de-energized loads. This case is similar to case 1, except the capacity of the line between F s/s and G s/s is decreased from 4.0 to 1.0. This means that the available power from *BAG9* decreased to 1.0; therefore the amount of available power becomes 3.0. As the total amount of de-energized loads is 4.0, the same as in case 1, the available power is insufficient to restore all the loads.

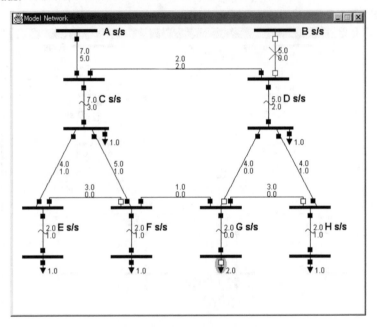

(a) Target network

NO	Station	Operation Contents	
1	Bs/s	B-D Line	Open
2	Ds/s	B-D Line	Open
3	Gs/s	D-G Line	Open
4	Ds/s	C-D Line	Close
5	Gs/s	Load	Cut
6	Gs/s	F-G Line	Close

(b) Switching sequence

Fig. 6.7. Case 2: Partial restoration of loads

Although one load at G s/s is unfortunately disconnected as shown in Fig. 6.7, this is the optimal solution under these conditions. This case requires four switching operations (open: 2, close: 2).

Double faults (Case3)

Next, a specific case is considered, where double faults occur at the same time. Fig. 6.8 shows the post-fault network where the lines between B s/s and D s/s and between C s/s and E s/s are tripped because of the assumed double faults. The simulation condition is similar to case 1, except the loads of BUS 12 and BUS 14 have been changed to 1.0 and 2.0 respectively. In this case, since there are two de-energized bus groups {D s/s, G s/s, H s/s} and {E s/s}, the restoration processes are started individually. Fig. 6.9 shows the simulation results. Six switching operations (open: 2, close: 4) are required to restore all loads.

It has been demonstrated through the obtained results that the proposed multi-agent system have found a sub-optimal target configuration in every case. The calculation time is within two seconds in all cases, including the time to form the operation sequence list.

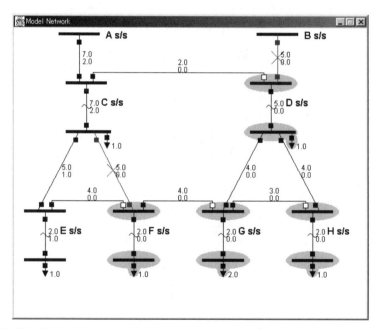

Fig. 6.8. Case 3: Post fault network

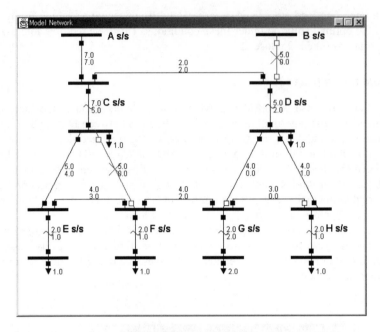

(a) Target network

NO	Station	Operation Contents	
1	Bs/s	B-D Line	Open
2	Ds/s	B-D Line	Open
3	Cs/s	C-F Line	Open
4	Fs/s	C-F Line	Open
5	Fs/s	E-F Line	Close
6	Gs/s	D-G Line	Open
7	Ds/s	C-D Line	Close
8	Gs/s	F-G Line	Close

(b) Switching sequence

Fig. 6.9. Case 3: Double faults

6.5 Conclusions

In this chapter, we have presented a multi-agent approach to power system restoration. The proposed multi-agent system consists of a number of *Bus Agents*

(*BAGs*) and a single *Facilitator Agent* (*FAG*). Several simple restoration strategies are embedded in every *BAG*, that communicate only with its neighboring *BAGs*, while the *FAG* acts to facilitate the decision process. While a *BAG* intends to conduct the local search, the *FAG* carries out the global search. The validity and effectiveness of the proposed multi-agent system have been demonstrated by applying it to a practical size model network. It is noted that the proposed multi-agent system can decide the target configuration and the switching sequence using local information only. This means that the proposed multi-agent system is a promising approach to more complex large-scale networks.

The next step of this work is to improve the multi-agent system's performance in order for it to cope with the difficulty of multiple faults. Future work should be directed to expanding this approach to handle bulk power restoration problems that involve the control of generators.

References

[1]   Simburger EJ, Hubert FJ (1981) Low Voltage Bulk Power System Restoration Simulation. IEEE Trans. on Power Apparatus & Systems, vol PAS-100, no 11, pp 4479-4484

[2]   Kafka RJ, Penders DR, Bouchey SH, Adibi MM (1981) System Restoration Plan Development for a Metropolitan Electric System. IEEE Trans. on Power Apparatus & Systems, vol PAS-100, pp 3703-3713

[3]   Kafka RJ, Penders DR, Bouchey SH, Adibi MM (1982) Role of Interactive and Control Computers in the development of a system restoration plan. IEEE Trans. on Power Apparatus & Systems, vol PAS-101, no 1, pp 43-52

[4]   McDermott TE, Drezga I, Broadwater RP (1999) A Heuristic Nonlinear Constructive Method for Distribution System Reconfiguration, IEEE Trans. on Power Systems, vol 14, no 2, pp 478-483

[5]   Gutierrez J, Staropolsky M, (1987) Policies for Restoration of a Power System. IEEE Trans. on Power Systems, vol PWRS-2, no 2, pp 436-442

[6]   Sakaguchi T, Matsumoto K (1983) Development of a Knowledge Based System for Power System Restoration. IEEE Trans. on Power Apparatus & Systems, vol PAS-102, no 2, pp 320-329

[7]   Kojima Y, Warashina S, Kato, M, Watanabe H (1989) The Development of Power System Restoration Method for a Bulk Power System by Applying Knowledge Engineering Techniques. IEEE Trans. on Power Systems, vol 4, no 3, pp 1228-1235

[8]   Hotta K, Nomura H, Takemoto H, Suzuki K, Nakamura S, Fukui S (1990) Implementation of a Real-Time Expert System for a Restoration Guide in a Dispatch Center. IEEE Trans. on Power Systems, vol 5, no 3, pp 1032-1038

[9]   Kirschen DS, Volkmann TL (1991) Guiding a Power System Restoration with an Expert System. IEEE Trans. on Power Systems, vol 6, no 2, pp 556-566

[10]   Simakura K, Inagaki J, Matsunoki Y, Ito M, Fukui S, Hori S (1992) A Knowledge-based Method for Making Restoration Plan of Bulk Power System. IEEE Trans. on Power Systems, vol 7, no 2, pp 914-920

[11]   Liou KL, Liu CC, Chu RF (1995) Tie Line Utilization During Power System Restoration. IEEE Trans. on Power Systems, vol 10, no 1, pp 192-199

[12] Nagata T, Sasaki H, Yokoyama R (1995) Power System Restoration by Joint Usage of Expert System and Mathematical Programming Approach. IEEE Trans. on Power Systems, vol 10, no 3, pp1473-1479

[13] Lee S, Lim S, Ahn B (1998) Service Restoration of Primary Distribution Systems Based on Fuzzy Evaluation of Multi-criteria. IEEE Trans. on Power Systems, vol 13, no 3, pp 1156-1163

[14] Talukdar S, Ramesh VC (1994) A Multi-agent Technique for Contingency Constrained Optimal Power Flows, IEEE Trans. on Power Systems, vol 9, no 2, 855-861

[15] Krishna V, Ramesh VC (1998) Intelligent Agent for Negotiation in Market Games, Part I: Model. IEEE Trans. on Power Systems, vol 13, no 3, pp 1103-1108

[16] Krishna V, Ramesh VC (1998) Intelligent Agent for Negotiation in Market Games, Part II: Application. IEEE Trans. on Power Systems, vol 13, no 3, pp 1109-1114

[17] Contereras J, Wu FF (1999) Coalition Formation in Transmission Expansion Planning. IEEE Trans. on Power Systems, vol 14, no 3, pp 1144-1149

[18] Nagata T, Sasaki H (2002) Multi-Agent Approach to Power System Restoration. IEEE Transaction on Power Systems, vol 17, no 2, pp 457-462

[19] Gamma E, Helm R, Johnson R, Vlissdes J (1995) Design Patterns: Elements of Reusable Object-oriented Software. Addison-Wesley

[20] Kuokka D, Harada L (1995) On Using KQML for Matchmaking. Proceedings of First International Conference on Multi-agent Systems, AAAI Press

7 Agent Technology Applied to the Protection of Power Systems

James S. Thorp[1], Xiaoru Wang[1], Kenneth M. Hopkinson[2], Denis Coury[3],
Renan Giovanini[3]

[1] School of Electrical Engineering, Cornell University, USA
[2] Department of Computer Science, Cornell University, USA
[3] Department of Electrical Engineering, School of Engineering of São Carlos,
University of São Paulo, Brazil

This chapter presents new explorations into the use of agent technology applied to the protection of power systems. We broadly define an agent to be an autonomous piece of software capable of interacting with its environment. Our investigation centers on the use of agents within an electric power grid using a private communication network, termed a Utility Intranet, based on Internet technology. Our goal is to create practical agents that can be examined in realistic situations. To facilitate this aim, a new platform has been developed, EPOCHS, which allows agents to operate in an environment that combines proven electric power electromagnetic and electromechanical transient simulators with a popular network communication simulation engine. We have developed a remote load shedding scheme, a back-up protection system and a differential protection method using agent technology and evaluated each case within the EPOCHS environment. Agent-based systems improve on traditional methods in each case. Results illustrating the performance and viability of agent-based protection system are presented.

7.1 Introduction

Power systems have undergone a great deal of change in the past decade. The United States electric power grid was deregulated in the late 1990's. As a result, it is increasingly operated with smaller generation reserves and more transmission congestion. If deregulation continues, it seems likely that power flows will be larger and will fluctuate to a greater extent in future years. Faster, more reliable, and better coordinated protection and stability control are even more critical under this new environment than they have been in the past. New methods are needed to meet this challenge.

Traditional protection systems rely upon standalone units that use local measurements as the basis for most decision making. Communication plays a very limited role in these legacy systems. A new era of decentralized control and efficiency demands has led to an environment demanding efficiency and reliability that pushes these legacy methods to their limits. The power system industry is beginning to recognize the benefits that communication could contribute towards greater system coordination, more rapid action, and increased correctness. The pervasiveness of the Internet has, at the same time, led to greater interest in using communication networks for improved protection systems. In particular, there has been an increasing interest within the power community in the use of networked agents to improve the reliability and efficiency of the electric power grid.

Despite the interest level in the electric power community, there have been limited opportunities to investigate the effects of agents on realistic systems due to the lack of a platform that could simultaneously simulate both electric power and communication elements. This chapter introduces agent studies utilizing EPOCHS, the **E**lectric **Po**wer and **C**ommunication **S**ynchronizing **S**imulator. EPOCHS is a new environment allowing users to create and test agent-based communications scenarios involving electric power transients.

We present simulation results of practical agents in realistic protection scenarios and show that they make improvements on traditional methods. Three scenarios were investigated: a backup protection system, a transmission line differential protection scheme, and a SPS control system. In all cases, agent-based approaches proved to be superior to more traditional alternatives.

7.2 Agent Technology in Power System Protection

7.2.1 Definition

The keyword 'agent' has been popularized both in and out of the artificial intelligence research community, but there is still no single universally accepted definition about what it means to be an agent. Software agents nearly always have the properties of autonomy and interaction. In addition, an agent may exhibit the properties of mobility, intelligence, adaptivity, and communication.

- *Autonomy* is the ability to take independent action.
- *Interaction* is the capacity to sense the surrounding environment and make changes to it.
- *Mobile agents* can literally travel from one location to another while maintaining their state and functionality. One can imagine a mobile agent moving through a network in search of certain data, or perhaps relocating to a new location with certain attributes. In contrast, *local agents* are placed by their owner and then remain in the same position throughout their computational lives. Both

types of agents normally have the ability to communicate with one another making mobility a lesser benefit than it might appear at first glance.

- *Intelligent agents* can learn from their environment. The term intelligent particular applies if the learning method involves artificial intelligence techniques, such as neural networks, fuzzy logic, etc.
- *Adaptivity* is the capacity to dynamically change behavior and receive updated parameter settings from the owner, from other agents participating in the same application, or from its own internal logic.
- *Communication* refers to the ability to exchange information between disparate entities. This information exchange could be as simple as a trip signal, or could be complex such as detailed information about a remote location.

In this chapter, the term agent will be used to refer to computer programs that are autonomous, interactive, and have the ability to communicate over a network. Agents may optionally also have any of the other attributes defined above.

7.2.2 Agent Architecture

Our agent proposal uses geographically distributed agents located in a number of *Intelligent Electronic Devices (IEDs)* as shown in Fig. 7.1. An IED is a hardware environment that has the necessary computational, communication, and other I/O capabilities needed to support a software agent. An IED can be loaded with agents that can perform control and/or protection functionality. These agent-based IEDs work in an autonomous manner where they interact both with their environment and with each other. An example of this might be digital relays where each one has its own thread of local control, but they perceive a more global scope of the system and act in response to their non-local environment by communicating with other agents either via *Local Area Networks (LANs)* or via *Wide Area Networks (WANs)*. For the purpose of this book chapter, an IED must be capable of receiving and supporting user-based agent software.

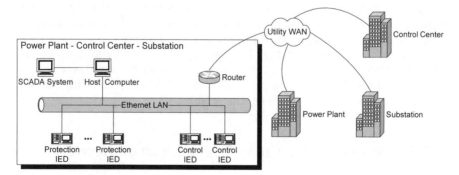

Fig. 7.1. Placements of the Agent-based IEDs within the Utility Intranet Infrastructure

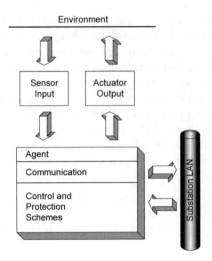

Fig. 7.2. The Structure of an Agent-based IED

Fig. 7.1 illustrates IED placement within a Utility Intranet infrastructure. Generally speaking, each agent makes decisions without the direct intervention of outside entities. Decision-making is made based on knowledge received from other agents as well as on local measurements. This autonomy adds resilience to agent-based methods because any one agent can continue to work properly despite failures in other parts of the system.

The agent-based IED's structure is depicted in Fig. 7.2. Agents within an IED perceive their environment through local sensors and act upon it through the IED's actuators. Examples of sensor inputs might include local measurements of the current, voltage, and breaker status. Actuator outputs might include breaker trip signals, adjusting transformer tap settings, and switching signals in capacitor banks. Agents might even interface with legacy systems such as *Supervisory Control and Data Acquisition (SCADA)* systems. The host computer shown in Fig. 7.1 could act as a bridge between the old and new systems in this type of situation. Internally, agents might be composed of many layers of functionality and control or may be contained in a single layer depending on the designer's specifications and implementation. As shown in Fig. 7.2, agents have the ability to communicate through a LAN in order to interact with other agents directly located on that same LAN, or can pass information along to the Utility WAN, i.e. the Utility Intranet, ultimately communicating with more remote IEDs.

7.3 The Structure of a Utility Communication Network

Networked computing systems are becoming increasingly prevalent in many areas and we believe that this growth will occur within electric utility systems as well. Technology is constantly changing, but we can make some educated guesses about what such systems will look like. First, the network systems will almost certainly be built from standard commercial off the shelf components. To do otherwise would be expensive both in terms of initial cost outlay and in system maintenance. This means that these networks will be based on Internet standards even if the systems remained independent of the global network conglomeration. We can already see hints that such changes are coming in recent standardization efforts such as the *Utility Communications Architecture (UCA)*. Based on this assumption, we have assumed the use of IP-based communication protocols in all of our agent scenarios.

Communications protocols are methods for transmitting data between networked computers. The most popular low-level protocol in use on the Internet is *IP (the Internet Protocol)*. IP packages a data unit into an entity known as a packet and makes a best-effort attempt to deliver that packet to the intended destination. Best-effort here means that if a problem occurs in the delivery attempt, perhaps because of network congestion, then the packet will be lost. IP is rarely used by itself. Higher-level transport protocols run on top of IP to provide added functionality.

The two most integral networking protocols in IP-based networks such as the Internet are the *Transmission Control Protocol (TCP)* and the *User Datagram Protocol (UDP)*. TCP is the more popular of the two. TCP's most notable contribution beyond IP's functionality is its ability to compensate for network losses by retransmitting data until it has been received by its intended destination. This is extremely valuable in many cases, but it can have drawbacks in time-sensitive situations since TCP delivers all data on a first-come first-serve manner. This means that information that is no longer relevant can stand in the way of current data if packet losses become too great. UDP, by contrast, only makes minor additions to IP. UDP's best-effort delivery guarantees can be a blessing in situations where data is only relevant for a limited time.

It is important to stress that IP is independent of the physical medium that data is transported over. The Network Interface layer of the IP protocol stack could be made up of any number of compatible networking standards. Fiber optic based Ethernet networking is becoming a popular choice, but the actual network chosen is unimportant from IP's perspective.

7.4 Developments of EPOCHS

7.4.1 Overview

This section introduces EPOCHS, the **E**lectric **Po**wer and **C**ommunication **S**ynchronizing Simulator. EPOCHS allows users to create and test agent-based communications scenarios involving electric power transients. The EPOCHS system is notable for its use of multiple high-quality commercial and research off the shelf simulators without modifying any simulation engine's core. All three of the simulation systems used by EPOCHS allow user extensions to their functionality and

EPOCHS takes advantage of these facilities. This is important because source code is rarely available for commercial off the shelf systems and yet they are being used increasingly for many areas of work due to their high quality and low cost. We feel that this trend will only increase over time and see no reason why simulation engines will be immune.

EPOCHS makes use of three major simulators: *PSCAD/EMTDC*, for electromagnetic transient simulation, *PSLF*, for electromechanical transient simulation, and *NS2*, to simulate network communication. All three simulation engines are well-known and are used frequently in their respective areas. EPOCHS also adds two components of its own. The *AgentHQ* presents a unified world view to all simulated agents and acts as a proxy for all inter-simulation information exchange. A *Run-Time Infrastructure (RTI)* is responsible for maintaining consistent simulation timing between all components and for routing information between them.

The individual simulation components are run in parallel. Frequent synchronization points are used to read and modify simulation states in order to keep EPOCHS scenarios running properly.

7.4.2 Related Work

There have been many other systems that combine simulators together in groups, or federations, using runtime infrastructures like the one employed by EPOCHS. Gary Bundy et al at the MITRE Corporation was the author of one of the first published combined systems when he created an aircraft combat simulation platform that was composed of both the *AWSIM* and *ModSAF combat simulators* [6]. Aircraft were able to exist either in an AWSIM or ModSAF simulation and those aircraft could be handed off between the simulators. The interaction and model synchronization took place using a management environment known as the *Aggregate Level Simulation Protocol (ALSP)*. The project was motivated by the fact that both AWSIM and ModSAF existed for combat simulations, but each included levels of detail that were missing in the other. It was decided that the most cost effective way to include all, or at least, most functionality of the two would be to couple the simulations together. The final product was not seamlessly interoperable, but workarounds were found for the most important features allowing for a successful final product.

The *Electric Power Research Institute (EPRI)* sponsored a joint effort between Honeywell and the University of Minnesota to create the *SEPIA (Simulator for Electric Power Industry Agents)* simulation engine. SEPIA allows researchers to study the effects of agents when used in power markets in the deregulated electric power grid. The project allows the investigation of many complex agent scenarios and includes an intuitive interface. Network communication was not included in the simulated parameters as its focus was on evaluating agents used for power auctions in the electric power grid [25].

Recent work centering on combining, or federating, simulation systems has focused on the use of the *High Level Architecture (HLA)*. HLA is an architecture that can be used to combine individual simulations, known as federates, together into combined simulators known as federations. The "glue" that holds these combinations together is a central component known as a Run-Time Infrastructure (RTI). The RTI routes all messages between simulation components and is responsible for making sure that simulation time is appropriately synchronized. The main drawbacks to this approach are firstly that it can be very difficult to modify existing simulations to conform to the HLA specification. This can be particularly true if the source code is not available as is the case in many commercial off the shelf (COTS) products. The military generally has source code available in military simulations, but the civilian populace generally makes heavy use of COTS software. Secondly, the HLA can be an inefficient means of combining federates together. The system is based on a publish-subscribe mechanism where any federate subscribing to another will receive all of its updated information whether it is needed or not.

There have been a number of very good papers describing how to adapt existing COTS simulations for use in the HLA or in similar environments. [32] [27] are two prominent examples. A few major projects have arisen where this has been done in practice without source code availability. Prominent examples are Sudra et al's [28] documented examples coupling COTS simulations for the purpose of manufacturing simulations and again in the area of automotive simulations together using their *Generic Runtime Infrastructure for Distributed Simulation (GRIDS)*. They use the concept of thin agents located at each server to do things such as relevance filtering, data value prediction, and data subscription management. An infrastructure composed of COTS software for manufacturing simulation is also being undertaken by McLean and Riddick as documented in [21].

A wide range of simulations have made use of agents. There are two main classes of agent-based simulators. The first class uses agents to act as a mechanism for combining simulation engines. The flexibility that agents provide can be used as filters to reduce inter-simulation traffic, for example. An example of this can be found in Wilson [34]. The second class of simulations uses agents to model entities within a simulated world. Lee makes use of a mix of continuous and discrete-time agents in an air-traffic control simulation in [18] using agents to represent, among other things, air-traffic controllers, aircraft, and air traffic generators. These are just two of many examples of both classes of agent-based simulations.

7.4.3 EPOCHS Simulation Description

PSCAD/EMTDC

PSCAD/EMTDC is used for electromagnetic transient simulation. EMTDC is a well-known electric power simulator. One of its main strengths is its ability to accurately simulate power system electromagnetic transients. That is, EMTDC models short-duration time-domain electric power responses. PSCAD is a graphical interface that is used to simplify the development of EMTDC scenarios. PSCAD is produced by the Manitoba HVDC Research Centre [20].

EMTDC simulates power system scenarios by solving a series of differential equations in a time stepped manner. It has very detailed electrical models making it well-suited to electromagnetic transient investigations. PSCAD/EMTDC requires that a step size be specified before execution begins, but the step can be changed as the simulation progresses.

PSLF

PSLF is used for electromechanical transient simulation. PSLF can simulate power systems with tens of thousands of nodes and is widely used by electric utilities to model electromechanical stability scenarios [12]. It models large systems in less detail than that available in PSCAD/EMTDC. It is a time-stepped environment similar to that offered in PSCAD/EMTDC. Once again, a time step size is specified before execution begins, but the step can be changed as the simulation progresses.

NS2

Network Simulator 2 (NS2) is an event-driven communication network simulator that was created through a joint effort between the University of California at Berkeley, Lawrence Berkeley Labs, the University of Southern California, and Xerox PARC. NS2 is a high-quality simulator that allows the creation of a wide variety of communications scenarios. It has built-in support for the mostly widely used network protocols and for the most popular research protocols. It is particularly valuable when studying TCP/IP behavior within IP-based networks [5].

AgentHQ

AgentHQ is a module that we developed to present a unified environment to our agents and acts as a proxy for those agents when interacting with other EPOCHS components.

Runtime Infrastructure (RTI)

The RTI is a module responsible for synchronization points and for routing messages between components. It was created as part of the EPOCHS development process.

7.5 Simulation Architecture

The logical relationship between the five EPOCHS simulator components is illustrated in Fig. 7.3.

7.5.1 The Run-Time Infrastructure

The RTI acts as the "glue" between all other components. It is responsible for simulation synchronization and for routing communication between EPOCHS components. The component synchronization employs a time-stepped model when running EPOCHS scenarios. Time steps are user-selectable and can be chosen depending on the granularity of a given case. Simulations can use a short time between synchronization points to compensate for the errors introduced by the decoupled simulation approach or can use larger time steps for faster execution.

Our synchronization approach allows us to use a convenient alternative to modifying the core source code. PSCAD/EMTDC, PSLF, and NS2 allow user-defined extensions. That is, a PSCAD/EMTDC scenario can include user-defined libraries that add equipment definitions using the C programming language that were not present in the original software. PSLF similarly allows user-defined equipment models using its proprietary interpreted EPCL language. NS2 has well-defined procedures for adding new communication protocols in C++ to the base simulation software. We have created our own equipment stubs whose sole purpose is to interact with EPOCHS's RTI at each synchronization point. Both PSCAD/EMTDC and PSLF use a user-modifiable fixed length between each of their time steps. Both systems allow users to modify the time step length at each interaction, however we chose to keep time steps consistent for easy interaction in our first EPOCHS release.

NS2 allows the use of timing events. Hence, we activate the agent headquarters synchronization point after a user-defined number of calculation loops have occurred inside the power system simulators. We also set a timer event that takes place after an equivalent amount of time in NS2. This process is simplified by the fact that PSCAD/EMTDC, PSLF, and NS2 are all single-threaded systems, so each system is effectively halted whenever a synchronization event takes place. Additional efforts would be required if that were not the case.

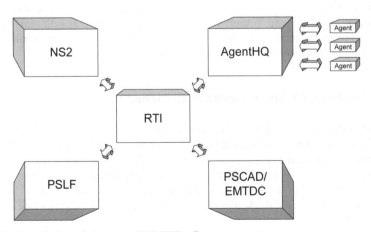

Fig. 7.3. The relationship between EPOCHS's five components

The synchronization between the various simulation components follows a simple algorithm. All systems are halted at time 0. At the beginning of any time step, the RTI waits for synchronization messages from both the power system simulator and NS2. Then, the RTI yields control to the AgentHQ. The AgentHQ passes the control on to the agents one by one until all have had a chance to execute. During this cycle, the agents are capable of sending communication messages and getting/setting power system variables. Once all agents are done, the AgentHQ returns control back to the RTI. Finally, the RTI notifies both NS2 and the power system simulator that the current time step is done. At this point, the two simulation engines run for an additional time step. Special attention must be paid to NS2. Messages may be received in between two synchronization points within NS2. If a message arrives, NS2 will immediately pass it along to the RTI bound for the AgentHQ. The AgentHQ will, in turn, pass the message on to the appropriate agent. The agent can process the message and send another in response. If the message requires power system state to be read or changed then that agent keeps the message in a queue until the next synchronization point occurs.

7.5.2 Electrical Component Subsystem

EPOCHS' electrical component subsystem is built on top of the power system simulator. The two power system simulators use a straightforward approach in their simulations. They run through an event loop at each time step where they calculate equipment values. We use a library in our simulations that adds a call to our user-defined component once per time step. The PSCAD/EMTDC component begins by reading in all user-accessible equipment values. The PSLF component has direct access to all internal variables making that step unnecessary. Next, the electrical component contacts the RTI and notifies it that the end of the time step has been reached. Agents can request equipment values or can set power values when they execute. At the end of an agent execution cycle, a finish message is sent from the RTI to the electrical components and the power component set any values that have changed in their simulations. The components relinquish control afterwards and execution continues. A flowchart of a power system component's execution can be found in Fig. 7.4.

7.5.3 The Network Communication Component

The network communication component is relatively simple. A new transport protocol is added to NS2 serving as its link to the RTI. A periodic call is added to the simulation script invoking the new protocol in order to halt execution and interact with the RTI once per time step. The length of the step can take on any value as long as it is the same as that used in the power system simulator. NS2 lacks the ability to automatically track message contents under most circumstances when using TCP/IP.

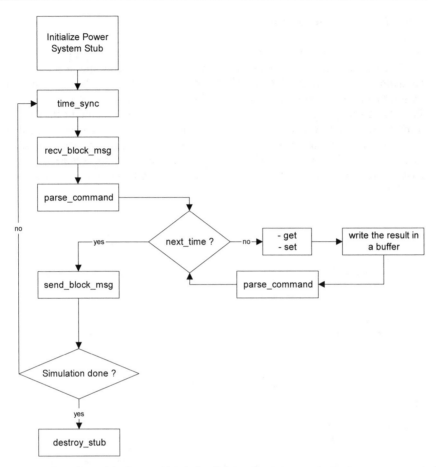

Fig. 7.4. A flowchart of the Power Simulation Subsystem

We use the TCPApp application, the exceptions to the rule, in order to keep track of this state on our behalf. All agents have *send* and *recv* methods that act as interaction points for network communication between the agents and the AgentHQ where they are passed to/from NS2 via the RTI. UDP, by contrast, does have the ability to transmit data and we took advantage of it adding our own layer of abstraction through a module we named UDPApp. This choice gives us a great deal of flexibility in terms of the communications protocol as we can easily send data through simple NS2 function calls. A flowchart of the NS2 component's execution can be found in Fig. 7.5.

7.5.4 The AgentHQ Subsystem

AgentHQ is triggered at each synchronization point and acts as a proxy between the agents, the network communications simulation, and the electric power simulators. At that time, the AgentHQ calls each of the agent's *action* methods giving them an opportunity to calculate their set of operations for the next time step. Agents may get or set values in power systems using method calls that pass through AgentHQ. AgentHQ waits for a return value and then passes the response back into the querying agent. Agents may also send messages to each other during their *action* activation time. Message calls can be made to AgentHQ and are passed into NS2 from there. A flowchart of the AgentHQ events can be found in Fig. 7.6.

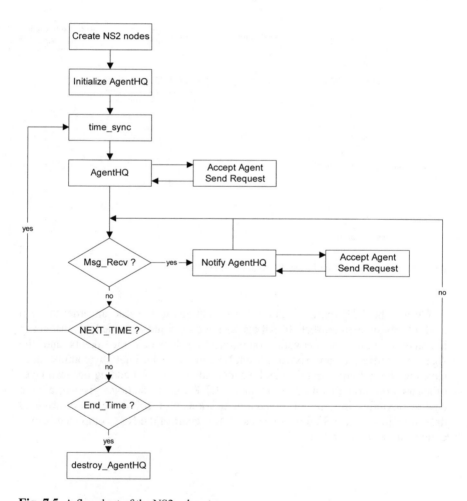

Fig. 7.5. A flowchart of the NS2 subsystem

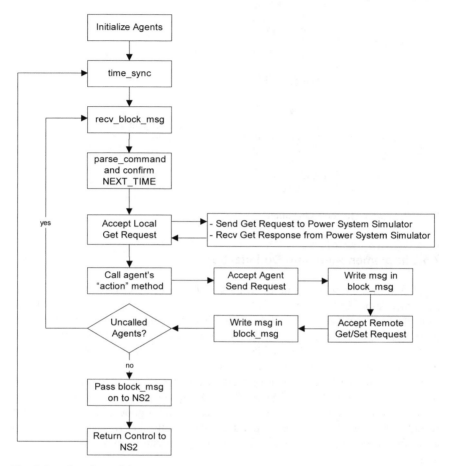

Fig. 7.6. A flowchart of the AgentHQ subsystem

If a message is received inside NS2 then it is passed on to the AgentHQ agent manager and is subsequently passed to the destination agent's recv method. The agent can perform any necessary message processing at that time and has the ability to send messages to other agents as well. A limitation of the coupled simulation system is that agents cannot immediately get or set power system variables upon message receipt because the electric power subsystem will be in the middle of an execution cycle at that time. An agent can compensate for this by placing such requests in a message queue to be dealt with at the beginning of the next synchronization period. A flowchart of this process can be found in Fig. 7.7.

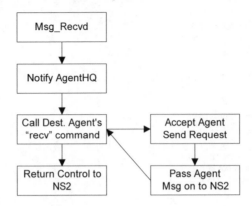

Fig. 7.7. Message Receive Operation

7.5.5 Implementation and Optimization

Fig. 7.8 shows the relationship between the different EPOCHS components in its implementation. In the current implementation, NS2, the RTI, the AgentHQ, and its corresponding Agents are all combined inside a single executable. Each component is logically separated within the source code and the RTI is still implemented as a protocol stub inside NS2. This combination was used purely to boost the performance of the simulation federation.

We have used EPOCHS to investigate a number of power system situations. None of them use PSLF and PSCAD/EMTDC at the same time. It is technically possible to do so, however we have not encountered a good example where this would be beneficial. As a result, we did not allow both PSCAD and PSLF to operate as federates at the same time, but it would only require a minor modification to do so.

7.5.6 Simulation Scripts

Each agent simulation takes three parts. The structure of the power system and its electrical parts must be laid out in a PSCAD/EMTDC or PSLF compatible file. The layout of the communications subsystem and the transport protocols used needs to be specified in an NS2 file. Finally, agent types and locations are added to the NS2 simulation script for use by the agent manager. If the agent component were decoupled from NS2 then this would result in a third simulation file that would be required by EPOCHS. This is one limitation inherent in combining multiple simulators together. Each must be initialized in its own way. The other major drawback is the extra communication and management overhead between the simulation components.

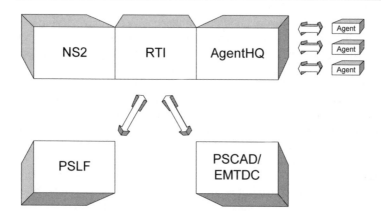

Fig. 7.8. The relationship between the EPOCHS Components in its Implementation

7.5.7 Summary

We feel that techniques like those used in EPOCHS will only become more common over time as the need to investigate interconnected situations crossing commercial simulation domain boundaries continues to increase.

7.6 Backup Protection Systems for Transmission Networks

7.6.1 Overview

Relay misoperations, such as failing to isolate a faulted line or incorrectly tripping a healthy line, are involved in major disturbances within power systems and can lead to catastrophic cascading events or even blackouts [23] [22]. Giving relays the ability to self-check and self-correct would be a significant addition at critical locations in the power system by helping to reduce the number of misoperations.

Backup protection is required to clear a fault when the primary protection fails. There are some drawbacks to traditional backup protection, which normally consists of step distance and step over-current relays. Backup relays have to be slow enough to coordinate with all related relays. Their isolated region is also much larger than necessary. In addition, zone 3 backup relays sometimes operate incorrectly due to heavy loads in abnormal system events, the most familiar cause of cascading outages.

Various backup protection systems based either on expert systems or adaptive architectures have been proposed in an attempt to correct for traditional relaying systems' shortcomings [26] [29] [30]. Both classes of methods require additional local and remote information. In the adaptive method, the zone 3 distance relay setting was adapted to ensure good performance under widely varying power system operating conditions. In a method utilizing expert systems, a single computer was placed in each substation where it was used to identify faulty lines and to trip the appropriate circuit breakers when faults were found.

A few proposals in the protection literature have alternatively used agents as components in a cooperative protection system [31], a distributed intelligence in a protection method [35], and an interactive, autonomous software process in an adaptive multi-terminal line protection scheme [11]. We propose an agent-based backup protection scheme where agents are embedded in conventional protection components to construct a relay IED. The relay agent searches for relevant information by communicating with other relay agents. Its purpose is to detect relay misoperations and breaker failures and perform backup protection with much better performance than can be expected from traditional methods.

7.6.2 The Architecture of the Agent Relay

Our backup protection proposal uses geographically distributed agents located in relay IEDs. This method improves on more traditional isolated component systems by allowing relays to take non-local information into account. In order to make these distributed systems work effectively, the agents must be capable of autonomously interacting with each other.

A natural way to combine legacy protection systems with modern distributed information infrastructures is to "wrap" the protection components with an agent layer enabling them to interact with other agents and hence other components [36]. Our agent relay IEDs, shown in Fig. 7.9, illustrate this approach.

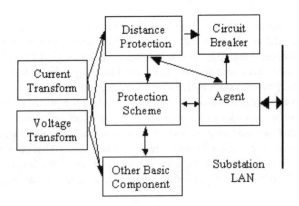

Fig. 7.9. The Relay IED Architecture

They each have their own thread of conventional control, but communicate with other agents and act in response to their environment. An agent makes decisions without the direct intervention of outside entities. This flexibility and autonomy adds reliability to the protection system because any given agent-based relay can continue to work properly despite failures in other parts of the protection system.

7.6.3 The Strategy Employed by the Agent-Based Backup Protection System

Agents monitor their corresponding relays and take corrective actions when failures occur in the protection system. An agent achieves this by searching for relevant information within its transmission area and by then taking action based on the rules given in Table 7.1. The information gathered generally includes the operational responses of conventional relays and circuit breakers' status. However, it may sometimes also include currents and voltages when using a differential method. A relay agent's transmission region generally consists of the transmission lines included in its primary and backup protection zones as well as those lines whose backup protection cover the current line. An agent can either receive the list of agents, which are in its transmission area and with which it will communicate, at initialization or it might learn this information through some type of network topology discovery algorithm.

As shown as in Table 7.1 Rule 1, when an agent observes a primary relay trip, the agent will check the status of its own zone 3 relays and will communicate with agents in its transmission area to find their relay status. The trip operation of the primary relay is considered to be correct if its zone 3 relay trips and the other terminal's primary relay trips, or incorrect when none of the zone 3 relays operates. Otherwise, the situation is uncertain and the agent will use a differential method to locate the fault. The primary relay trip is correct if a fault is found in its primary protection zone or is incorrect if it is not.

If the correctness of a primary relay trip has been confirmed, the agent takes note that a breaker trip signal has been sent. It then waits until the breaker clearing time has passed. If the breaker is still closed at that time then the agent will send trip message to all other agents in the area surrounding the fault in order to clear it in the smallest practicable region. These agents will send trip signals to their breakers and will then initiate their respective breaker failure supervision routines. This process continues until the fault is cleared. In this way, the agent system provides fast backup protection with high selectivity.

Based on rule 5 in Table 7.1, even though both the primary and local zone 3 relays fail to respond to a fault, the remote zone 3 relays that responded to the fault will send messages to the agent whose relay failed to locate the fault initiating the use of a differential method. A trip signal will be sent afterwards to open the faulted line.

Table 7.1. Rules for agent behavior

Rules	Conditions	IF	Then	Action
1	A primary relay trips	No zone 3 relays trip in its transmission area	The primary relay trip is incorrect	Block the breaker trip if possible
		Its zone 3 relay trips and the other terminal's primary relay trips	The primary relay trip is correct	Monitor for breaker failure
		Other	Uncertain 1	Go to rule 3
2	A zone 3 relay trips	The correct primary relay trip in its backup zone has been confirmed by rules 1 or 3	The zone 3 relay trip is correct	No action
		Other	Uncertain 2	Go to rule 4 and send notification messages to the agents in its transmission area
3	Uncertain 1 (Differential method used)	A fault is identified in its primary zone	The primary relay trip is correct	Monitor for breaker failure
		No fault is found in its primary zone	The primary relay trip is incorrect	Block the breaker trip if possible
4	Uncertain 2 (Differential method used)	A fault is identified in the primary zone	The zone 3 relay is correct but the primary relay has failed	Trip the breaker and monitor for breaker failure
		A fault is identified in the remote backup zone by another agent	The zone 3 relay is correct	No action
		Other	The zone 3 relay trip is incorrect	Block the breaker trip
5	The agent gets a notification message, but its relay has not tripped (Differential method used)	A fault is identified in the primary zone	The primary and zone 3 relays have failed	Trip the breaker and monitor for breaker failure
		A fault is identified in the remote backup zone by another agent	The zone 3 relay has failed	No action
		Other	The zone 3 relay behavior is correct	No action

Therefore, better performance is achieved compared with the bigger isolated region and longer fault clearing time of traditional remote zone 3 relay actions. The agent-based protection system has many other benefits over traditional systems. For example, zone 3 relays can be monitored allowing corrections to prevent false breaker trips. This correction has the potential to greatly reduce the number of false trips in heavily loaded cases.

In summary, our agent-based backup protection system has the ability to self-check, self-correct, and rapidly reacts while achieving highly selective fault regions backup functions when either primary protection or circuit breakers fail. To illustrate the agent's advantages, we will look at an example involving transmission line MN in Fig. 7.10.

7.6.4 Simulation results

We created two experiments to illustrate the advantages of our agent-based backup protection system. Experiments were run using PSCAD/EMTDC in the EPOCHS environment. We designed the agents and their supporting infrastructure using an object-oriented approach. All experiments revolved around the 400 kV example system shown in Fig. 7.10.

As described in 7.3, there are two main IP transport protocols. TCP provides reliable transport while UDP makes a best-effort attempt to deliver a message, but message delivery is not guaranteed. We have chosen to focus our initial backup protection communication investigations using the UDP protocol for its nearly universal support and low latency despite its lack of delivery guarantees. We rely on traditional backup protection behavior in the event of communication failure.

The simulation was simplified by only considering three-phase faults and by using the zone 1 distance relay as primary protection. The transmission network is equipped with an agent-based backup protection system. Any agent can communicate with any other agent. The admittances of current transformers, voltage transformers, and anti-aliasing filters are included and the DFT is used to obtain inputs needed for mho distance relays. Both distance and time delay relays have been placed into operation with traditional settings. The tests include an incorrect operation of a zone 1 relay and a breaker failure.

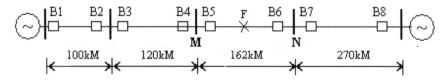

Fig. 7.10. The example power system

Agent communication was simplified to the most basic level. We assumed that there is a 1 ms delay between each bus and that there is no network congestion. Agent messages consisted of either requests for voltage, current, and relay status, breaker set requests, or replies to those requests.

Case I: Primary Relay Misoperation

In the first experiment, a zone1 relay misoperation occurs at time 0.2 seconds. The simulation results are shown in Fig. 7.11. It can be seen that there is a block signal sent by the relay agent 5 to breaker 5 near time 0.2 seconds, preventing the breaker trip initiated by the incorrect operation of relay zone 1. The resulting phase A current can be seen in Fig. 7.11 (c). Had the agents not been in the system, the phase A current would look like Fig. 7.11 (d).

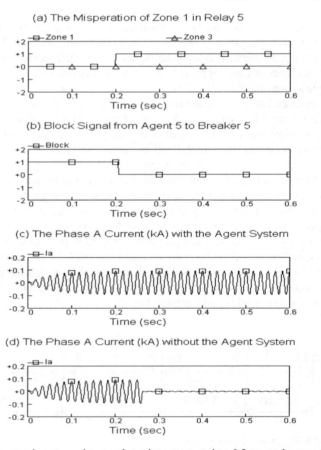

Fig. 7.11. In this experiment, a primary relay misoperates at time 0.2 as can be seen in (a). The agent-based backup protection system responds to the failure (b) and does not open the breaker resulting in (c). The traditional system responds by opening the breaker (d)

Case II: Break Failures Occur

In the second experiment, a fault is located at the midpoint of line MN and is scheduled to occur at 0.3 seconds as can be seen in Fig. 7.12 (d). Moreover, after a fault occurs at 0.3 seconds, relay 5's agent confirms the zone 1 relay's correct operation. It perceives a breaker 5 failure when the setting time is longer than the breaker clearing time. The agent will react by sending a trip message to agent 4 (see Fig. 7.12 (b)), which initiates a signal to breaker 4 to clear the fault. However, breaker 4 is faulty leading to another message from agent 4 to agent 3 as shown in Fig. 7.12 (c). Agent 3 finally sends a signal to breaker 3 to clear the fault. The phase A current values for the entire experiment can be seen in Fig. 7.12 (d). However, much more time is needed for the fault to be cleared in remote zone 3 in the same situation without the agent system, as shown in Fig. 7.13.

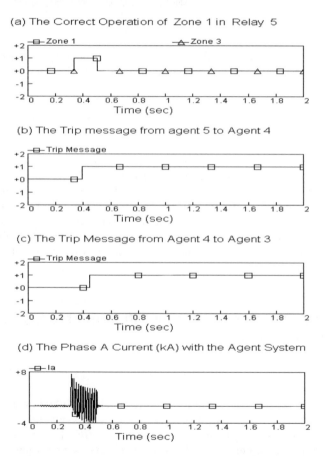

Fig. 7.12. Both breakers 4 and 5 fail when a fault occurs at time 0.3 seconds. The agent-based backup protection system clears the fault in just a little additional time beyond twice the breaker clearing time

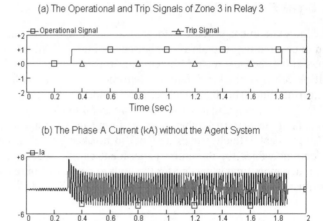

Fig. 7.13. This is the same situation as in Fig. 7.12 without the agent-based protection system. It takes a traditional zone 3 backup system more than 1.5 seconds to clear the fault remotely

7.6.5 Summary

In this section we developed a backup protection system that improves on the traditional trip zone while simultaneously decreasing clearing times. We have introduced a distributed systems architecture based on interactive autonomous agents. The agents embedded in each relay IED communicate with other relay agents in order to check the status of relays, locate faulted lines, and to minimize the isolated region when primary protection systems or circuit breakers fail. Experimental results show that this technique has promise and we hope to refine our ideas in upcoming work.

7.7 An Agent Based Current Differential Relay for Transmission Lines

7.7.1 Overview

This section describes an agent-based differential relay utilizing a Utility Intranet for the primary protection of transmission lines. It is shown that agents located in digital relays can use communication to ensure correct performance over a wide variety of operation conditions. Very few publications concerning the application of agents to the field of protection have been reported in the literature. Tomita et al [31] proposed a cooperative protection system utilizing agents. Relay agents were constructed. These agents cooperated to enhance traditional primary, backup and adaptive protection. In [10], Coury et al discuss the use of differential agent relays for protection of multi-terminal lines and results illustrating the performance of the proposed agent-based differential method are presented.

7.7.2 Related Work

Current differential relays are widely used for the efficient protection of buses, generators, and power transformers. When applied to transmission lines, problems arise due to the distance between the line ends making a communication channel necessary. The channels generally used for this purpose are power line carrier, microwave, communication cable, or even fiber optics [13]. The differential technique relies on the fact that, under normal conditions, the sum of the currents entering each phase of a transmission line from all connected terminals is equal to the charging current for that phase. If this sum exceeds a preset threshold, a fault may exist inside the protection zone.

Additional problems may arise due to the improper synchronization of data between the two ends of the protected line when utilizing a digital communication link [16] due to mismatched data sets. Some recent publications concerning digital differential protection of transmission lines try to address these problems. Redfern et al [24] proposed digital current differential protection using the limited data transfer capabilities of a digital data voice-frequency grade communication channel with satisfactory results. Li et al [19] proposed using Global Positioning Systems (GPS) for data synchronization. The results show precisely synchronized differential protection and the effect of current transformer saturation on the protection sensitivity.

7.7.3 The Differential Protection Agent Architecture

Fig. 7.14 illustrates the differential protection agent. It should be emphasized that, since the communication network is available, differential logic is the main method used for transmission line protection. The internal logic can be understood to be divided into a basic scheme for differential current calculation augmented by additional features.

Basic Scheme for the Current Differential Calculation

After passing through *Capacitive Voltage Transformers (CVTs)* and *Current Transformers (CTs)*, as shown in Fig. 7.14, the three phase voltage and current signals are digitized at 1.0 kHz. The digital signals are processed using their Fourier voltage and current components. Perfect line transposition is assumed. This assumption does not add significantly to the method's inaccuracy.

Initially in Fig. 7.14, a one cycle *Discrete Fourier Transform (DFT)* is used to extract the 60 Hz fundamental current and voltage components at the line terminals. Following that, a line PI model is used for the shunt current compensation.

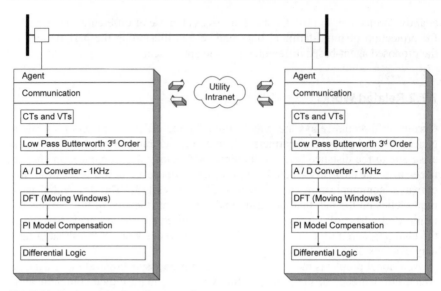

Fig. 7.14. Agent mechanism for protection purposes

At relay R this compensation is performed using Eq. (7.1):

$$\underline{I}' = C\underline{V} + D\underline{I} \qquad (7.1)$$

where: V, I are the fundamental components, and C, D are line constants.

The proposed method transmits current phasors rather than sampled values. In order to accomplish this, a one-cycle moving window was used. The 60 Hz phasors, or complex numbers, of the currents were transmitted through the network at 1 KHz. The largest advantage in transmitting phasor values is that if samples are dropped in the transmission process, this will not invalidate the differential logic calculation at the two ends of the line. After sampling the signal at 1 KHz using anti-aliasing filters, the currents are transmitted through the communication network in opposite directions and the differential logic is calculated at both ends of the line. If an internal fault is detected, circuit breakers associated with the agent relays will isolate the protected transmission line.

Additional Features

Some additional features for the agent-based relay are presented below:

- The agent-based differential protection system must work concurrently with a traditional distance protection scheme. This means that the agent relays must constantly exchange power system status and equipment failure information.
- The agent-based scheme must be able to detect equipment failure (CVTs, CTs, circuit breakers, etc) based on measurements. In the case of local equipment failure, the agent-based relays must collect data from other pieces of equipment through the communication network. Similarly, if a circuit breaker fails, a trip

signal must be sent through the communication network to either local or remote ends in order to isolate a fault.

- The agent-based scheme must be able to detect communication channel failures as well as GPS (Global Positioning System) signal failures. In the first case, a switch to traditional distance protection must be made. In the second case, other software-based methods for signal synchronization base must be relied upon.

7.7.4 Simulation Results Using EPOCHS

The Power System and the Communication Topology Utilized

In order to test the proposed scheme, the agent-based current differential relays were simulated acting in a 400 kV power system simulated by PSCAD/EMTDC. The power system consists of a tee configuration connected to a two-ended 100-km long transmission line as depicted in Fig. 7.15. The communication links used by the agents are also depicted. Every link in this scheme has a 2 MB/s capacity and 1 ms traversal time.

Results of the Agent Based Current Differential Relays

A set of 8 tests was created for the proposed differential protection system. During these tests, it was assumed that all faults occurred at 0.6 seconds and were of a fixed type (three phase to ground). During these tests, two parameters were varied: fault location and fault resistance. The values for these parameters were:

Fault locations:

- (1) : 50 Km from relay 1;
- (2) : 15 Km from relay 2 (between relays 1 and 2);
- (3) : 40 Km from relay 3;
- (4) : 50 Km from relay 5.

Fault resistances:

- (1) : 0.001 Ω;
- (2) : 300 Ω.

A detailed fault detection process is shown in the first two tables. In Table 2, a three phase to ground fault located 50 Km from relay 1 with a fault resistance of 0 Ω is presented, while in Table 3, a three phase to ground fault located 50 Km from relay 5 with a fault resistance of 300 Ω is presented.

In these tables, the first column shows the time that the differential calculation was performed while the second column shows the time when the current samples where taken. It is possible to infer from the table that the messages take roughly 2 ms to arrive from the remote relays to the local relay and be processed. The last two columns show the differential values calculated by the relays for phase A. Only one column is shown for relays 1, 2, and 3 since they share the same data.

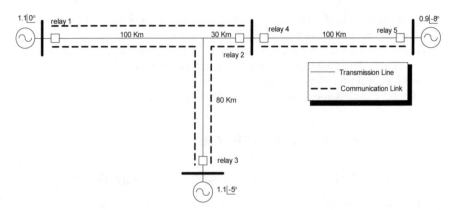

Fig. 7.15. The power system utilized for tests

The same observation is valid for relays 4 and 5. It must be pointed out that these values correspond to the magnitude of the differential current values calculated after the CT transformation and digital signal processing. Similar results were obtained for phases B and C.

Table 7.2 shows the correct operation for relays 1, 2, and 3. They were able to identify a fault in their primary zone in 6 ms. As we expected, relays 4 and 5 didn't identify any fault inside their primary zone.

Table 7.3 depicts the correct operation for relays 4 and 5. They identified a fault in their primary zone in 7 ms. Once again, the relays outside the primary zone protection for this fault didn't trip.

The last table presented in this section summarizes all tests performed on the power system. It shows how long each fault took to be identified. As can be seen, all relays worked corrected, clearing the fault within a half-cycle time frame.

Table 7.2. Three phase to ground fault located 50 Km from relay 1 with a fault resistance of 0.001 Ω

time (s)	currents sampling time (s)	Idiff relays 1,2,3 (A)	Idiff relays 4,5 (A)
0.598	0.596	0.0012	0.0008
0.599	0.597	0.0008	0.0003
0.600	0.598	0.0008	0.0003
0.601	0.599	0.0007	0.0003
0.602	0.600	0.0007	0.0004
0.603	0.601	0.0008	0.0005
0.604	0.602	0.0006	0.0004
0.605	0.603	0.0114	0.0015
0.606	0.604	0.2951	0.0083
0.607	0.605	1.6726	0.0079
0.608	0.606	4.8086	0.0080
-	-	-	-
0.630	0.628	49.8040	0.0106

Table 7.3. Three phase to ground fault located 50 Km from relay 5 with a fault resistance of 300 Ω

time (s)	currents sampling time (s)	Idiff relays 1,2,3 (A)	Idiff relays 4,5 (A)
0.598	0.596	0.0012	0.0008
0.599	0.597	0.0008	0.0003
0.600	0.598	0.0008	0.0003
0.601	0.599	0.0007	0.0003
0.602	0.600	0.0007	0.0004
0.603	0.601	0.0008	0.0005
0.604	0.602	0.0007	0.0004
0.605	0.603	0.0008	0.0011
0.606	0.604	0.0015	0.0317
0.607	0.605	0.0006	0.1640
0.608	0.606	0.0016	0.3589
-	-	-	-
0.630	0.628	0.0020	1.9757

Table 7.4. Summary of the tests performed in the power system

fault location	fault resistance (Ω)	relays 1,2,3 trip instant (s)	relays 4,5 trip instant (s)
1	0.001	0.606	-
1	300	0.607	-
2	0.001	0.606	-
2	300	0.607	-
3	0.001	0.606	-
3	300	0.607	-
4	0.001	-	0.606
4	300	-	0.607

7.7.5 Summary

This section introduced an agent-based current differential relay for use with a communication network. Results illustrating the performance of the proposed method were presented. It can be concluded that the new method utilizing agent technology offers some additions features beyond those present in traditional alternatives. Equipment failure detection and the capability to collect data from any point within a power system are both notable improvements provided by this approach. Finally, our proposed method illustrates that relays can use communication networks to perform the differential protection scheme for transmission lines. The results provided were very accurate and fast, showing the benefits of this technique.

7.8 Special Protection Systems

7.8.1 Overview

Power system instability usually involves large areas and results in dire consequences. Loss of generator synchronism for a single or group of generators with respect to another group of generators is a transient instability that can result in costly power blackouts. Stability problems typically involve disturbances such as the loss of generation, loads, or tie lines. These disturbances stimulate power system electromechanical dynamics and the system response typically includes deviations in frequencies, voltages, and generator phase angles. *Special Protection Schemes (SPS)* are mechanisms generally designed to counteract power system instability. The most common SPS schemes take action through generation rejection and load shedding, according to a recent report on industrial experiences with SPSs throughout the world [1].

Generation rejection is an effective and economic option for system stabilization in the event that either a generating plant's outgoing lines or an area's tie lines are tripped in order to prevent the remaining transmission lines from overloading or violating transfer limits. Generation rejection schemes are heavily dependent on a system's load flow transmission patterns, the type of fault encountered, and the fault's clearing time. SPSs generally detect a situation calling for generation rejection by monitor a connecting transmission line's status. In the simplest type, generation rejection is performed based on preset values, but the drawback to this method is that the amount of generation to be rejected has to be set for the worst possible contingency. This will typically result in the rejection of more generation than is necessary. An improvement can be made by basing decisions both on line status and on the type of disturbance so that the level of generation to be rejected with a single line to ground fault is less than it would be with a multiphase fault. Several approaches have been proposed for more accurately determining the amount of generation to reject based on transient stability analysis [9]. However, these methods are difficult to apply in real-time applications due to their high computational burden.

Remote load shedding is a remedial action scheme that is conceptually similar to generation rejection, but is applied to the load portion of the system. It is also used to reduce power mismatches in systems deficient in generation. This is done to assure that the system meets certain minimum operating standards and to facilitate the rapid restoration of customer load. Remote load shedding schemes can be used to maintain a system's frequency within a preset range. The WECC's policy is to keep its system frequency above 59.5 Hz [33]. This level reasonably assures that the WECC will have the ability to restore their frequency to 60 Hz by utilizing available reserves. Hydro-Quebec uses 58.5 Hz as their minimum frequency level [8]. One of the goals of remote load shedding schemes is to minimize dependence on *underfrequency load shedding (UFLS)*. The main advantage of remote load shedding over standard UFLS is that it both reduces the load trip delays and lowers the load shedding requirements. According to the WECC Southern Island load tripping plan for the loss of the California-Oregon Intertie (COI) and subsequent northeast/southeast separation, the remote load shedding delay is assumed to be 20 cycles. In this WECC plan, remote load shedding amounts are set to 10%-20% less than the standard UFLS value where the relay is set to a higher level of 59.5 Hz. For some contingencies, the frequency drops so rapidly that no UFLS response other than fast remote load shedding can save the system. The load shedding amount is the most significant parameter in remote load shedding schemes. The Southern Island uses three possible load shedding levels. Their level selection is done based both on their imported and stored energy. Canada's load shedding amount is also adjusted as a function of the system status, loading, and configuration. These systems are an improvement over simple fixed SPS systems, but their limited flexibility is still far from ideal. Faster and more accurate estimation methods are still needed to determine the proper load shedding values in real-time applications.

In this section, we have developed an agent-based generation rejection and load shedding SPS. It is designed to react to a severe fault in a major EHV transmission line in the case where another outage has already taken place in the same corridor. Our goal is to prevent instability and to preserve the system's integrity within a safe operating frequency range. We will present the algorithm for determining the proper amount of load that should be shed to hold the system's frequency above a preset level based on wide area measurements. The proposed scheme has been tested with a modified version of the IEEE 50 generator test case and simulated with the EPOCHS simulation environment. The results are very promising.

7.8.2 Algorithms for Frequency Stability Control

There have been several methods proposed for frequency stability control based both on local measurements [3] [4] [7] and on wide area measurements [17]. We make contributions to the estimation of disturbance size when confronted with electromechanical dynamics during an emergency. Our method allows us to compute the steady state frequency after a disturbance and the amount of load shedding required to maintain a preset frequency level taking the effect of generator governors into account.

A Dynamic Model for Frequency

The swing equation that governs generator motion is approximately [2]

$$2H\dot{\omega} = P_m - P_e \tag{7.2}$$

where H is the generator's inertia constant in seconds and is often normalized to the rated capacity S_{Bmach} of that machine, P_m and P_e are respectively mechanical and electrical power normalized to that same VA base, and ω is the electrical angular velocity or frequency in per unit values. For a system having N machines in operation we have

$$2\sum_{i=1}^{N} H_i \dot{\omega}_i = \sum_{i=1}^{N} \left(P_{mi} - P_{ei} \right) \tag{7.3}$$

where H_i, P_{mi} and P_{ei} are based on system base VA $S_{Bsystem}$.

We define the system inertia constant H_{system} in relation to system angular frequency ω_{system}, system mechanical power $P_{msystem}$, and system electrical power $P_{esystem}$ as

$$H_{system} = \sum_{i=1}^{N} H_i = \sum_{i=1}^{N} H_{machi} \left(S_{Bmachi} / S_{Bsystemi} \right) \tag{7.4}$$

$$\omega_{system} = \sum_{i=1}^{N} \left(H_i \omega_i \right) \bigg/ \sum_{i=1}^{N} H_i \tag{7.5}$$

$$P_{msystem} = \sum_{i=1}^{N} P_{mi} \tag{7.6}$$

$$P_{esystem} = \sum_{i=1}^{N} P_{ei} \tag{7.7}$$

We define the system accelerating power

$$P_a = P_{msystem} - P_{esystem} \tag{7.8}$$

Given equations (7.4)-(7.8), we can rewrite (7.3) as

$$2H_{system} \dot{\omega}_{system} = P_a \tag{7.9}$$

A disturbance, such as a sudden change of load or generation, will induce the dynamic frequency behavior in the system based on the above equation. Turbine governor mechanisms and the system's load sensitivity play a significant role in the final outcome.

The Estimation of the System's Disturbance Size

In the first instant following a disturbance of generation loss in the system, the accelerating power consists of the disturbance P_d and the change in electrical power demand ΔP_e due to the variation of frequency and voltage

$$P_a = P_d - \Delta P_e \left(\omega_{0^+} - \omega_{0^-}, u_{0^+} - u_{0^-} \right) \tag{7.10}$$

where $_{0^-}$ and $_{0^+}$ respectively denote the time immediately before and after the disturbance. ΔP_e describes the change in power loss and load demand as follows

$$\Delta P_e = \Delta P_{loss} + \sum_{j=1}^{M} \Delta P_{lj} \qquad (7.11)$$

where M is the number of loads.

The system accelerating power P_a can be obtained with (7.9) based on wide area measurements where H_{system} is calculated by (7.4) using the generators' operating status, and $\dot{\omega}_{system}$ is computed using the initial samples after the disturbance of the generators' electrical angular velocities at intervals of length h according to the following equation.

$$\dot{\omega}_{system}(k) \approx \frac{\omega_{system}(k) - \omega_{system}(k-1)}{h} \qquad (7.12)$$

This change in electrical power demand ΔP_e can also be obtained using wide area electrical power measurements both immediately before and immediately after the disturbance.

$$\Delta P_e = \sum_{i=1}^{N} P_{eo^+} - \sum_{i=1}^{N} P_{e0^-} \qquad (7.13)$$

Thus, the disturbance size can be estimated by equation (7.10). i.e.

$$P_d = P_a + \Delta P_e \qquad (7.14)$$

Estimation of the Required Amount of Load Shedding

Let us assume that we are given the size of a disturbance. We wish to find the required load shedding amount p_{shed} to ensure that the system's steady state angular frequency will stay above a preset value ω_{set} in the near term, for example within 30 seconds after the disturbance.

At the steady-state point, denoted by ∞, after a disturbance, $\dot{\omega} = 0$, the power balance equation can be derived from (7.9)

$$P_d + \Delta P_m \left(\omega_{set} - \omega_{0^-} \right) - \Delta P_e \left(\omega_{set} - \omega_{0^-}, u_\infty - u_{0^-} \right) + P_{shed} = 0 \qquad (7.15)$$

where ΔP_m is the power contribution due to governor action, and ΔP_e is the change in electrical power demand due to the voltage and frequency variation between the values before the disturbance and those in steady state.

The mechanical power increase for each generator due to governor action is

$$\Delta P_{mi} = -\Delta\omega / R_i \qquad (7.16)$$

Parameter R represents speed regulation. Methods neglecting turbine limits when calculating the rise in mechanical power will yield unsatisfactory estimations. In fact, when there is a frequency decrease due to a drop in generation or increase in load, only generating units with generation remaining will respond. Normally, only a fraction of spinning reserve is instantly available. The load that can be picked up by a thermal unit is limited by its turbine. About 10% of a turbine's rated output can be initially picked up without causing heat damage followed by a slow increase of about 2% per unit [2]. Therefore, we have to add a constraint to (7.16) when computing power output due to governor action. The following constraints are based on a modified method presented by M. E. Connolly and mentioned in the discussion section of reference [3] :

$$\Delta P_{mi} \leq S_{Bmachi} - P_{mi0^+} \qquad (7.17a)$$

$$\Delta P_{mi} \leq K_i S_{Bmachi} \qquad (7.17b)$$

where P_{mi0^+} is the turbine power output immediately after the disturbance. $S_{Bmachi} - P_{mi0^+}$ represents the remaining generation. $K_i S_{Bi}$ is the generations' short-term response limit.

P_{mi0^+} can be obtained by the following equation derived from (7.2) using the measurements of generator's electrical power output and angular frequency derivatives immediately after the disturbance.

$$P_{mi0+} = 2H_i \dot{\omega}_{i0^+} + P_{ei0^+} \qquad (7.18)$$

If the system is in steady-state before the disturbance, P_{mi0+} can be obtained by simply measuring P_{ei0^-} because $P_{mi0^+} = P_{mi0^-}$ and $P_{mi0^-} = P_{ei0^-}$.

By setting $p_{shed} = 0$ and $\omega_\infty = \omega_{set}$, the system's steady state frequency ω_∞ without a loading shedding can be solved iteratively using (7.15)-(7.18). We can also determine whether remote load shedding is required by checking if the predicted steady state frequency is within an acceptable range.

7.8.3 The System Studied and Agent-Based SPS Scheme

The System Studied

Our experimental simulations make use of IEEE's 145-bus 50-generator system, shown in Fig. 7.16 [14], with some modification. The first system modification made was to add a 500 kV line from bus 1 to bus 25. The added line increased the number of system branches from 453 to 454. Our intent was to create a case that necessitated the use of a SPS in order maintain system stability. Power systems can generally sustain the loss of a single tie line, but require action with the loss of a second if the line is not cleared quickly enough. Next, the total system capacity was reduced to 30,050.00 MW. This lowered system capacity made the 4,277 MW power flow along the main 500 kV transmission corridor much more important than it was in the old system. The percentage of admittance load was also reduced to 5.02% and the remaining 94.98% was set as constant active and reactive power. The six machines located at buses 93, 104, 105, 106, 110, and 111 are all represented with two-axis machine models and equipped with IEEE type AC4 exciters. All other generators are represented by classical machine models. All generators are equipped with basic steam turbines and employ governors with a 5% droop setting [12]. System analysis is performed after governors have responded, but before new load reference set points can be established by the AGC subsystem.

Description of the SPS Architecture

The SPS is required to provide rapid and reliable action. The communication requirements are different from that of traditional SCADA systems due to the need for fast information updates and rapid response to commands dictating generation rejection and load shedding. In our example, the main SPS agent is located at bus 1, a 500 kV substation. The main agent identifies extreme contingencies such as the loss of two lines and performs both generation rejection with preset units and load shedding based on real-time measurements. The generators that can be rejected are selected based on simulation studies. The main SPS agent communicates with other agents to gather data values including generators' connection status, active power outputs, angular frequencies and frequency derivatives. It also communicates with other agents located at major system or load buses to collect voltage and frequency measurements as well as the load available for shedding.

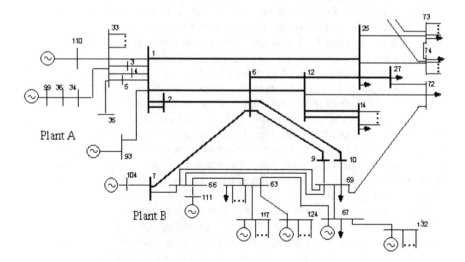

Fig. 7.16. One line diagram of major 500kV lines, generators and loads in the 50-generator system

Generator agents are located at power plants where they send their measurements to main SPS agents when requested to do so. Generator agents also reject generation when ordered to do so by those same main SPS agents. Load agents are mainly located at distribution substations. These agents shed load upon receiving an order from SPS agents. They also locally perform underfrequency load shedding. Underfrequency load shedding might occur, for example, if the frequency reached a threshold value of 57~58.5Hz after a remote load shedding scheme with a preset frequency of 58.8 Hz fail to hold the system frequency above 58.5 Hz.

7.8.4 Simulation Results

Initially, the 500 kV transmission lines from buses 1 to 6, 1 to 25, and the two lines from 1 to 2 carry respectively 722.2 MW, 1,106.5 MW, and 2×361.1 MW. A trip on the 1-6 line causes the system to operate under stressed conditions. The power in lines 1-25 and 1-2 respectively increase to 1,384MW and 2×515.7 MW. If a three phase fault occurs near bus 1 when line 1-6 has already tripped then the critical fault clearing time for opening line 1-25 is around 0.065 second.

We created a communication network with a node at each bus and an edge between nodes wherever one or more transmission lines existed in the case file. We set a conservative lower bound our transmission link delays by setting them to 0.500 seconds. We set each link's bandwidth to 150 Megabits per second so that no packets would be lost due to network congestion. Per-link loss rates were given values of 0 first and then the experiment was rerun with a loss rate of 0.05. The simulation with 5% link loss rates was performed in order to get a very basic idea of how the SPS system would react to situations with imperfect communication. All communication was sent using the UDP transport protocol.

In experiments, a fault occurs on the 1-25 line at time 0, the fault is cleared at time 0.07 seconds, and a trip command is sent to generator 93 at time 0.10 seconds. Because the fault is cleared after the critical fault clearing time, the system becomes transiently unstable and one group of 17 generators loses synchronism with respect to another group of 33 generators. Fig. 7.17 shows the rotor angle of a representative sample of generators with respect to the generator at slack bus 145.

The main SPS agent at bus 1 recognizes the situation and begins to communicate with other system agents to gather data values including generators' connection status, active power outputs, and angular. In the mean time, the main agent issues a generation rejection order to the agent at bus 93. Bus 93 is chosen here based on an off-line simulation study. Generation rejection keeps the system stable, as shown in Fig.7.18 (a), but the system frequency drops to around 57.5 Hz, as shown in Fig. 7.18 (b).

The main agent detects the "disturbance" created by generation rejection at bus 93. It begins to estimate the disturbance size and the size is found to be -1,862 MW. The main agent find that there is a 2,090 MW generation remaining and predicts that the steady state system frequency after this disturbance is 57.45 Hz. This predicted frequency value, as shown as in Fig. 7.18 (b), is very close to the simulation results.

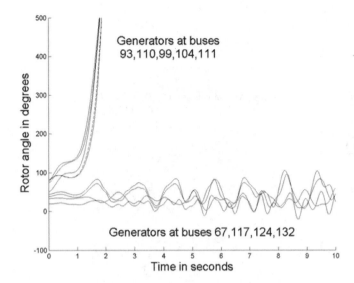

Fig. 7.17. Rotor angles with transient instability induced by a disturbance

Because the preset frequency is 58.8 Hz in our system, remote load shedding is required. The load shedding amount to maintain the system frequency above that preset level is estimated as 886 MW. We shed this total load at buses 14, 25, 27, 63, and 69 with the same percentage allocated to each load. That is, respectively, 85.97 MW, 293.29 MW, 182.25 MW, 157.16 MW and 167.92 MW. The system's frequency response, as shown in Fig 7.19, demonstrates the merit of our approach.

Fig 7.20 shows that the SPS operation with a 5% loss rate per link is different from that without communication losses. The reason for this difference might be that the wide area information used to calculate the disturbance size and the load shedding amount are different for the two cases. That is, if too many communication packets are lost then there can be a large time difference between the beginning of the disturbance and the data point where that disturbance was computed.

7.8.5 Summary

We have proposed a novel frequency prediction and control algorithm based on wide area measurements. We made use of this algorithm in an agent-based generation rejection and load shedding scheme that is able to keep a system transiently stable and is able to maintain the system's frequency above a preset level when there is an emergency tie line loss. Experimental results show the accuracy and usefulness of the proposed approach under light communication loss rates. However, inaccuracies are introduced under heavier communication traffic levels. There are many promising areas of future work related to what we have done.

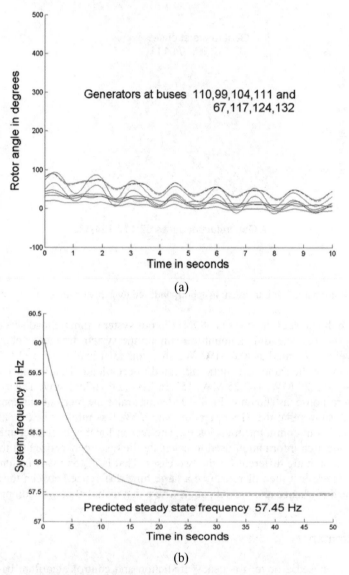

(a)

(b)

Fig. 7.18. (a) The system remains stable after generation rejection (b) The system undergoes an unacceptable frequency decrease

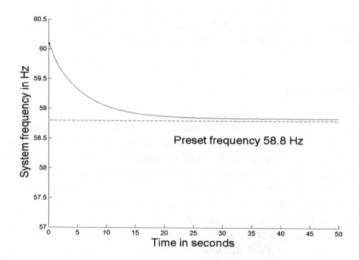

Fig. 7.19. Remote load shedding maintains the system frequency above a preset level

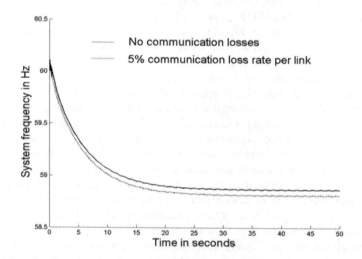

Fig. 7.20. Comparison of the SPS operation with different levels of loss rate

We use a very simple method to estimate the system's load sensitivity, and that method works well in our experiments using a system with 94.98% constant power load and 5.02% admittance load. A system that contained more complex load types would require a correspondingly more complex load model for accurate results. Our work also shows that agents are beneficial when applied to one SPS system. Agent-oriented approaches have the potential to be even more effective when multiple SPS systems overlap. Uncoordinated SPS systems have the potential to counteract each other and wreak havoc on a power system under unfavorable conditions. We believe that a good agent-based system can prevent such scenarios from happening by allowing SPSs to coordinate their actions.

7.9 Conclusions

In this chapter, we have described the use of agents for power systems protection. We first defined the concept of agents by describing their main characteristics. One of the most fundamental features of an agent is its ability to communicate. We believe that utilities will construct their own private communication networks in the future to be used in applications like those outlined in this chapter. Our experiments assumed that this private network, termed the Utility Intranet, was positioned on top of the electric power grid. A Utility Intranet will almost certainly use standard Internet components for IP communication. For that reason, all experiments were based on TCP/IP and UDP/IP, the standard IP network transport protocols.

To investigate agent-based scenarios, we created EPOCHS, a combined electric power and communication simulator designed for that purpose. EPOCHS makes use of PSLF, PSCAD/EMTDC, and NS2. In all cases, new functionality was added to those simulators through the use of custom modules, separate from the original source code. We presented important agent-based systems in the areas of protection and demonstrated their utility using the EPOCHS testbed. In the first case, a backup protection system was developed. This system improved protection area selectivity and decreased fault clearing times demonstrating the advantages of cooperative systems over traditional standalone alternatives. In the second case, a differential protection system was created. We showed that agents could greatly increase system protection performance by exchanging basic information within their local zones through network communication. Finally, a generation rejection and load shedding SPS system was created that could prevent system instability and to preserve the system's integrity within a safe operating frequency range.

The agents developed in this chapter don't represent a break from traditional protection systems. Agent-based systems necessarily rely on communication networks to add functionality beyond what can be achieved by traditional methods. Agent-based systems can incorporate traditional functionality within them to compensate for communication failures. In this way, agent-based systems enhance system security, but techniques that depend on communication are not relied upon as the sole protection mechanism. It is still far too early to predict the future of agent-based protection systems.

There are many issues that must be dealt with in agent-based protection system. Communication security, proper communication protocol selection, and the appropriate range of applications for agent-based methods are just a few of many issues that must be dealt with before agents will be ready for use in real systems. Nonetheless, the tools and techniques examined in this chapter demonstrate early promises for improved protection systems in the future.

References

[1] Anderson P, LeReverend BK (1996) Industry Experience with Special Protection Schemes. IEEE Transactions on Power Systems, vol 11

[2] Anderson PM (1993) Power System Control and Stability. IEEE Express

[3] Anderson PM, Mirheydar M (1990) Low-Order System Frequency Response Model. IEEE Transactions on Power Systems, vol 5

[4] Anderson PM, Mirheydar M (1992) An Adaptive Method for Setting Underfrequency Load Shedding Relays. IEEE Transactions on Power Systems, vol 7

[5] Breslau L, Estrin D, Fall K, Floyd S, Heidermann J, Helmy A, Huang P, McCanne S, Varadhan K, Xu Y, Yu H (2000) Advances in Network Simulation. IEEE Computer, vol 33, pp 59-67

[6] Bundy GN, Seidel DW, King BC, Burke CD (1996) An Experiment in Simulation Interoperability. Presented at Winter Simulation Conference,

[7] Chuvychin VN, Gurov NS, Venkata SS, Brown RE (1996) An Adaptive Approach to Load Shedding Scheme and Spinning Reserve Control During Underfrequency Conditions. IEEE Transactions on Power Systems, vol 4

[8] CIGRE System Protection Schemes in Power Networks. SCTF 38.02.19

[9] CIGRE (1999) Advanced Angle Stability Controls. CIGRE TF 38.02.17

[10] Coury D, Thorp JS, Hopkinson KM, Birman KP (2002) An Agent Based Current Differential Relay for Use with a Utility Intranet. IEEE Transactions on Power Delivery, vol 17, pp 47-53

[11] Coury DV, Thorp JS, Hopkinson KM, Birman KP (2000) Agent Technology Applied to Adaptative Relay Setting for Multi-Terminal Lines. Presented at IEEE PES Summer Meeting, Seattle, USA

[12] General   Electric   (2003)   PSLF   Manual.   [Online].   Available: http://www.gepower.com/dhtml/corporate/en_us/assets/software_solns/prod/pslf.jsp

[13] Horowitz SH, Phadke AG (1996) Power System Relaying. John Wiley and Sons

[14] IEEE Committee Report (1992) Transient Stability Test System for Direct Stability Methods. IEEE Transactions on Power Systems vol. 1

[15] Kundur P (1994) Power System Stability and Control. McGraw Hill, New York, NY

[16] Kwong WS (1988) Current Differential Protection Goes Digital. Journal of Modern Power Systems, vol 1, pp 53-56

[17] Larsson M, Christian R (2002) Predictive Frequency Stability Control Based on Wide-Area Phasor Measurements. Presented at IEEE Power Engineering Society Summer Meeting

[18] Lee S, Pritchett A, Goldman D (2001) Hybrid Agent-based Simulation for Analyzing the National Airspace System. Presented at Winter Simulation Conference

[19] Li HY, Southern EP, Crossley PA, Potts S, Pickering SDA, Caunce BRJ, Weller GC (1997) A New Type of Differential Feeder Relay Using the Global Positioning System for Data Synchronization. IEEE Transactions on Power Delivery, vol. 12, pp 1090-1097

[20] Manitoba HVDC Research Centre (1998) PSCAD/EMTDC Manual Getting Started. Winnipeg, Manitoba, Canada

[21] McLean C, Riddick F (2001) The IMS Mission Architecture for Distributed Manufacturing Simulation. Presented at Summer Computer Simulation Conference

[22] Phadke AG, Horowitz SH, Thorp JS (1999) Aspects of Power System Protection in the Post-Restructuring Era. Presented at Hawaii International Conference on System Sciences

[23] Phadke AG, Thorp JS (1996) Expose Hidden Failures to Prevent Cascading Outages. IEEE Computer Applications in Power, vol 1, pp 20-24

[24] Redfern MA, Chiwaya AAW (1994) A New Approach to Digital Current Differential Protection for Low and Medium Voltage Feeder Circuit using a Digital Voice-Frequency Grade Communication Channel. IEEE Transaction on Power Delivery, vol 9, pp 1352-1358

[25] Samad T, Harp S, Wollenberg B, Morton B, Pires L, Brignonne S (1999) Siimulation of Complex Systems for the Power Industry with Adaptive Agents. EPRI, Palo Alto, CA TR112816

[26] Stedall B, Moore P, Johns A, Goody J, Burt M (1996) An Investigation into the Use of Adaptive Setting Techniques for Improved Distance Back-up Protection. IEEE Transactions on Power Delivery, vol 11, pp 757-762

[27] Strabburger S (1999) On the HLA-based Coupling of Simulation Tools. Presented at European Simulation Multiconference

[28] Sudra R, Taylor SJE, Janahan T (2000) Distributed Supply Chain Simulation in GRIDS. Presented at Winter Simulation Conference

[29] Tan JC, Crossley PA, Hall I, Farrell J, Gale P (2001) Intelligent Wide Area Back-up Protection and its Role in Enhancing Transmission Network Reliability. Presented at IEE International Conference on Developments in Power System Protection

[30] Tan JC, Crossley PA, Kirschen D, Goody J, Downes JA (2000) An Expert System for the Back-up Protection of a Transmission Network. IEEE Transactions on Power Delivery, vol 15, pp 508-514

[31] Tomita Y, Fukui C, Kudo H, Koda J, Yabe K (1998) A Cooperative Protection System with an Agent Model. IEEE Transactions on Power Delivery, vol 13, pp 1060-1066

[32] Tucci M, Revetria R (2001) Different Approaches in Making Simulation Languages Compliant with HLA Specifications. Presented at Summer Computer Simulation Conference

[33] WECC (2003) Southern Island Load Tripping Plan. [Online]. Available: http://www.wecc.biz/documents/politcy/wscc_southern_island_tripping_plan_approved_july_97.pdf

[34] Wilson LF, Burroughs D, Sucharitaves J, Kumar A (2000) An Agent-based Framework for Linking Distributed Simulations. Presented at Winter Simulation Conference

[35] Wong SK, Kalam A (1997) An Agent Approach to Designing Protection Systems. Presented at IEE Conference Publication, no 434

[36] Wooldring M, Ciancarina P (2000) Agent-oriented Software Engineering: The State of the Art. Presented at AOSE, Limerick, Ireland

8 Dynamic Output Compensation between Selected Channels in Power Systems

Serge Lefebvre

Institut de Recherche d'Hydro-Québec (IREQ), Varennes, Québec, Canada

This chapter proposes a framework for controlling power systems involving two types of control actions: first swing and damping controls. The problem is addressed with control agents having each a specific objective. The agents concur to provide system-wide control. A Wide-Area Measurement System provides the information necessary at each agent using its own representation of the system. First swing controls act as special protection systems in multiple time scales and channels. Damping controls rely on low order dynamic multiple input single output MISO controllers between selected input and output channels at each control agent. The design objectives of damping controls are to mimic the relevant dynamics of linear state feedback. Structural constraints are accounted by a design based on the best measurements available at each agent.

8.1 Introduction

The power system is operated so that it can continue to supply all users even when a major component fails. This is first contingency operation. Security of the power system is achieved when it can withstand a set of credible contingencies. Power systems become vulnerable when the designed protection/control system cannot maintain stability of the power system in unanticipated and complicated situations. Threats, from the sources of vulnerability, can be reduced by installation of a defense system that decreases the probability and the severity of occurrence. For example, [1] presents defense plans to maintain reliability of the Hydro-Québec's transmission system in the face of extreme contingencies. Defense plans apart, many utilities use special protection schemes (SPS) to maintain stability, reduce line overloads, maximize transfer capability, and provide voltage support.

Decision processes in a power system range from the very rapid ones preprogrammed into protective control equipment to the very slow ones associated with expansion planning. The information, required for the decisions, is encapsulated in models, or perhaps in operating policies. The information may also processed immediately, as a controller input or as a signal to system operators.

Largely, linear methods are used design power system controls, and then nonlinear system simulations, over a wide range of operating conditions, serve to tune the controls or derive SPS actions. Most stability controls seek to influence generator swing activity, or the voltage support that generators provide to the transmission network. The dynamics under control rarely occur at bandwidths above 2 Hz. The actuator, and the control law driving it, may very well have bandwidths higher than this. For example, HVDC links and Static Var Systems SVS of the Hydro-Québec network may have controllable responses up to 20–25 Hz [2].

Reference [3] presents several design methods used for power systems. Most practical power system controls rely on dynamic output feedback, often Single Input Single Output SISO controllers. Often only local variables enter the controllers and this has been successful for relatively simple oscillation patterns. Requiring that all modulation signals be local can make controller siting a difficult robustness issue [4]. Several control theories offer opportunities to significantly improve classical power system controls. However, the availability of states and system size limit the practical implementation in power systems. Fortunately, not all remote signals are equally vital to controller performance, or equally difficult to transmit reliably and without excessive delays.

During the last decade, the interest has been on managing system-wide power system dynamics through the deployment of WAMS measurement systems for acquisition of dynamic data and to resort to global controllers. Real-time data is useful since there can be some unusual topological configurations in the system. For example, after the ice storm of January 1998, Hydro-Québec experienced long-term transmission line outages because of downed transmission towers. An intelligent autonomous system would be able provide on-line operator support. Violations after a disturbance would be mitigated either by SPS agents, or by corrective actions of system operators when time allows.

Distribution equipment vendors have started to address distribution system automation with distributed intelligent agents [5]. However, comprehensive monitoring and control of a large power system is a long step beyond the monitoring of local devices or even regional performance. In the transmission world, so far, disturbance recording, disturbance analysis and calibration of dynamic system models have been the outcome. For example, Hydro-Québec planners use a snap-shot from the state estimator and the topology processor to determine system conditions at the time of an event for more detailed off-line analysis. Operating rules, SPS and damping controls are then derived from a range of studies [6].

This chapter proposes a framework for power system control with agents having each a specific objective. The framework is amenable to autonomy of the agents. The agents concur to provide system-wide control addressing global objectives. A Wide-Area Measurement System provides the information necessary at each agent using its own representation of the system.

8.2 Framework

In the context of the electric power industry, the scale of the power system can be anywhere from thousands to tens of thousands of nodes, with an array of interconnections between the nodes. The supervision and management subsystems are naturally distributed among these nodes. There are a large number of software and hardware tools to work together to meet the overall operation objective and this will grow as technology allows [7].

Information sharing capability is one novelty since, nowadays, lower hierarchical levels have no information access to the same or a higher level. An agent is an entity that performs autonomous actions based on information; it may rely on other agents to acquire or share information to achieve its goals. In principle, an agent on a level is able to access all information from all agents, if necessary. Agents can be developed for stability and security control that can provide an array of services such as reactive power generation, power flow control, harmonic compensation, voltage regulation, or dynamic control over the frequency and voltage. A large set of decision criteria is necessary for the agents such that they can make control decisions that will ultimately improve the power quality and reliability of the electric grid.

Fig. 8.1 illustrates the framework in which three regional networks form a large power system. The higher control level provides set points and some models to the lower level; it receives signals from all regions. The secondary level coordinates control actions within a region and may exchange information with its neighboring secondary controller. Each region uses several local area controllers. The resulting structure is decentralized, but global and coordinated. Each controller operates as a stand-alone unit. In certain circumstances, there is a requirement for the individual controllers to interact with one another. Once models are constructed [8]-[10], control schemes can tune system performance and stability, as control agents have access to time-tagged feedback. This does not imply that individual controllers will, nor should, all have access to the entire information. It is crucial to extract the goal-significant information from measurements and to communicate this to agents, whose structure may or not be pre-determined.

The information communicated is not restricted to be purely local information but it may not be the same at all agents. For example, process variables, such as regional speed or angle, need to be shared between various agents. Each entity involved has only a local view of the entire scope, while the entire system should react in such a way that all entities are closely coordinated.

In Fig. 8.1, the local agents react based on the "fuzzy and global" objectives of the regional agents. These in turn react to a demand of the global controller, which has its own picture of the system. The demand may be to reduce the load in their region by a specified amount, to limit the regional angle acceleration, to control the power exchange with a neighbor, or to control the voltage profile.

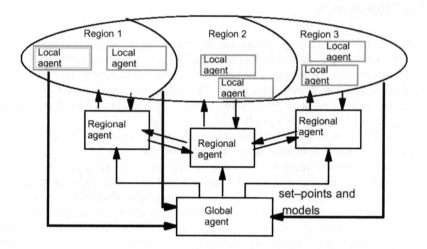

Fig. 8.1. A generic view of power system control

Fig. 8.2 provides more information on a region with its zones. It is up to the regional agents to organize actions to achieve the desired global objective and their own objectives. The agents communicate information such as control objectives, control and security performance, topology, models, SCADA and high-speed measurements. Not all information has the same priority and accuracy.

8.3 SPS Agents

This section addresses SPS agents and illustrate their behavior.

8.3.1 Principles

Fig. 8.3 illustrates a regional SPS agent whose main objective is to prevent loss of synchronism. To achieve this objective, control actions must be strong and fast and may include a defense plan in case of emergencies. Regional decisions are based on high-speed real-time regional key measurements, such as the angles. The upper level coordinates the actions and set the longer terms objectives.

The system can be decomposed in N interacting regions. Each region, for example region i containing N_i generators, is described by simple swing equations. These involve the variables Y_{ikm} and Θ_{ikm}, respectively the amplitude and phase angle nodal admittance between generators k and m in region i, E_{im} the internal voltage of machine m in region i, as well as M_{ik} and D_{ik} the inertia and damping coefficients, P_{mik} the mechanical power driving the machine and P_{eik} its electrical power.

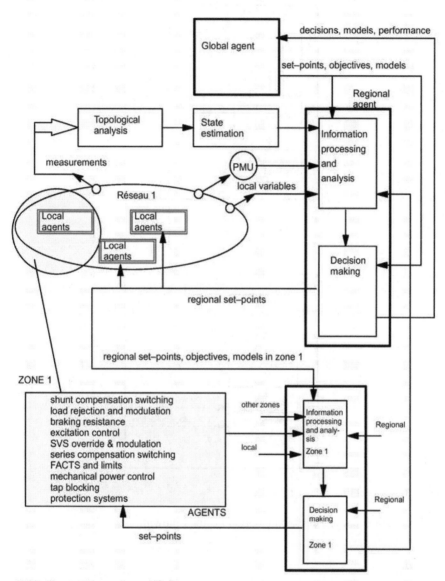

PMU: Phasor Measurement Unit

Fig. 8.2. Entities involved in power system control

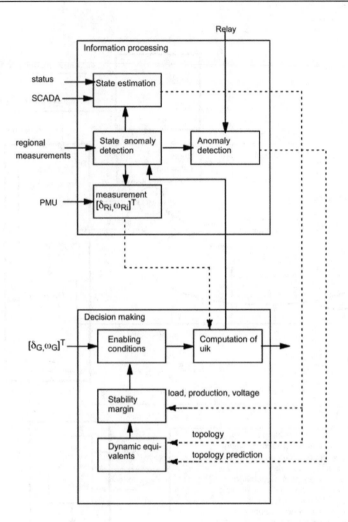

Fig. 8.3. Regional first swing and SPS agents

The set of equations describing region i is the following

$$\forall k = 0,1,2,N_i$$

$$\frac{d\delta_{ik}}{dt}(t) = \omega_{ik}(t)$$

$$\frac{d\omega_{ik}}{dt}(t) = M_{ik}^{-1}(P_{mik} - D_{ik}\omega_{ik}(t) - P_{eik}(\delta_i, u_i, t) + u_{ik})$$

$$(8.1)$$

$$P_{eik}(\delta_i, u_i, t)\Big|_{u=0} = E_{ik}\sum_{m=1}^{N_i} E_{im}Y_{ikm}\cos(\Theta_{ikm} - \delta_{ik} - \delta_{im})$$

Assume that the fast acting SPS control variable u_{ik} directly affects the mechanical power. The state variables $[\delta_i, \omega_i]$ represent the machines angle and speed in region i. The regional state is the weighted average of its machine states. With M_{Ri} the inertias of the ith region :

$$\delta_{Ri}(t) = (\frac{1}{M_{Ri}}) \sum_{k=1}^{N_i} (M_{ik} \delta_{ik}(t))$$

(8.2)

$$\omega_{Ri}(t) = (\frac{1}{M_{Ri}}) \sum_{k=1}^{N_i} (M_{ik} \omega_{ik}(t))$$

The global state $[\delta_G, \omega_G]$ is the weighted average of the regional states.

$$\delta_G(t) = (\frac{1}{M_G}) \sum_{i=1}^{N} (M_{Ri} \delta_{Ri}(t))$$

(8.3)

$$\omega_G(t) = (\frac{1}{M_G}) \sum_{i=1}^{N} (M_{Ri} \omega_{Ri}(t))$$

These measurements provide information on the system for the first swing and SPS actions. Line angle measurements are also useful. Critical lines signature are also excellent indicators of zones and regions behavior. The lines to monitor may be identified from off-line investigations of the likely topologies and operating conditions. Other measurements are used, such as breaker statuses.

There is a gradation of the actions due to the requirement of speed of response and of information handling. Without explicit knowledge of the system, initially the agents derive their actions from purely local measurements. The zone measurements may be added, followed by regional and global ones, where regional and global agents are put to contribution.

For example, in the examples to come, the actions on machines k-m-o in the ith region are in three time-frames.

Local action:

$$u_{ik} = -A \, s_k q_k \Delta\omega_{ik}$$
$$s_k = 1, if \ \Delta\omega_{ik} \succ 0; \ else = 0$$
$$A = 1, \ if \ t \prec t_0; else \ A = e^{-\beta(t-t_0)} \ with \ \beta \succ 0$$

(8.4)

Local and zone action

$$u_{ik} = -A \, s_k q_k \Delta\omega_{ik} - B_k \Delta\omega_{km}$$
$$u_{im} = -A \, s_m q_m \Delta\omega_{im} - B_m \Delta\omega_{mk}$$
$$B_{k/m} = 0, \ if \ t \prec t_1 \ or \ t \succ t_2$$

(8.5)

Local, zone and regional actions

$$u_{ik} = -A_k s_k q_k \Delta\omega_{ik} - B_k \Delta\omega_{km} - C_k \Delta\omega_{Ri}$$

$$u_{im} = -A_m s_m q_m \Delta\omega_{im} - B_m \Delta\omega_{mk} - C_m \Delta\omega_{Ri}$$

$$u_{io} = -A_o s_o q_o \Delta\omega_{io} - C_o \Delta\omega_{Ri}$$

$$C_{k/m/o} = 0, \text{ if } t \prec t_2$$

(8.6)

The key is to select the appropriate channels for the measurements and the actions.

8.3.2 Example

The example uses the 16-machines 68-bus model of the New-England NPCC network with loads modeled as constant impedance. There are two regions shown in Fig. 8.4. Region 1 represents the 39-bus New-England benchmark. Reference [11] describes the coherent zones. The example is a simple illustration of the processes involved.

Fig. 8.5 is the phase-plane representation of the zone R_{1-67} including generating stations 6 and 7. The event is a three-phase fault near generator 6 cleared by tripping the line between generators 6 and 7. The first phase-plane illustrates the zone behavior with respect to the entire region R_1, while the other shows the zone behavior with respect to the entire network. In both cases, the plots indicate a coherent zone with respect to the rest of the system. The residual line linking the generating station 7 to the system, line 23-24, is monitored. The slope of the separation angle on this line, as illustrated in Fig. 8.6, is a good indication of the event severity inside the zone. It drives the damping actions that serve in Fig. 8.7. Without control, the fault is critical. Restricting the control to be purely local improves the first swing but it is not effective in restoring steady state quickly. When the agent uses zone measurements, it becomes more effective, and even more when the regional coordination comes in, around 5.5 sec.

8.4 Damping Agents

Damping controls use low order dynamic MISO controllers. The set of MISO controllers yields a MIMO controller between selected pairs of input channels and output channels, noted **MMISO** controller for Multiple MISO controllers. This goes beyond parameter tuning and derivation of supplementary control signals. From the viewpoint of large-scale systems theory, many relevant results exist for structural modeling and decentralized controls in the sense of local output feedback structure. Regarding decentralized controllers for general linear time-invariant interconnected components, existence theorems exist but it is not clear how to obtain the controls. Given this, we propose to adopt a structural modeling of the power system, and then, to use a sequential design approach in which the following items are considered: agent and interface modeling error, performance robustness, order of the controllers, and availability of measurements.

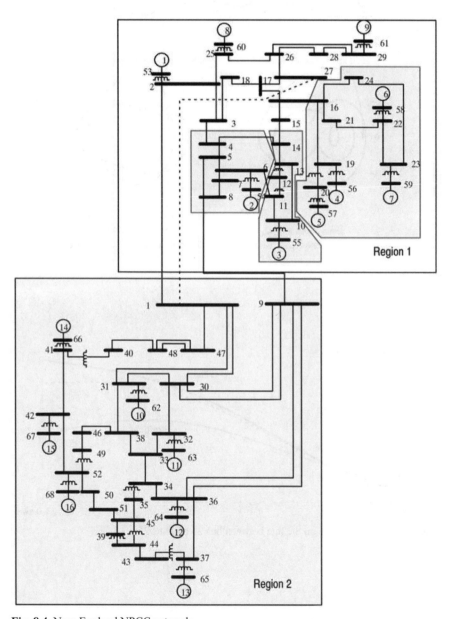

Fig. 8.4. New-England NPCC network

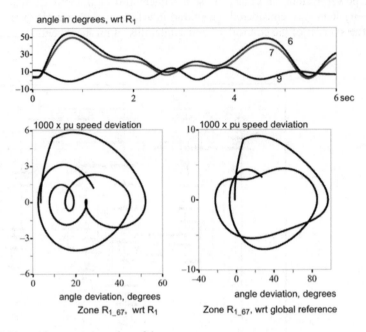

Fig. 8.5. Phase plane representation of the event

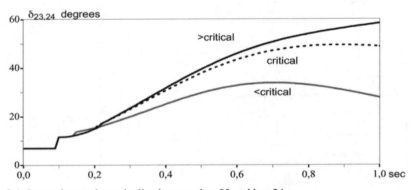

Fig. 8.6. Separation angle on the line between bus 23 and bus 24

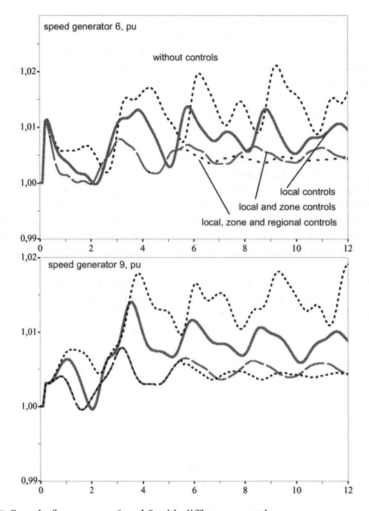

Fig. 8.7. Speed of generators 6 and 9 with different controls

8.4.1 Linear System Model as a Connection of Agents

The system is modeled as interconnected agents. The state-space representation of the ith agent is

$$\frac{d}{dt}x^{(i)} = A^{(i)}x^{(i)} + B_1^{(i)}a^{(i)} + B_2^{(i)}u^{(i)}$$

$$z^{(i)} = C_1^{(i)}x^{(i)} + D_1^{(i)}a^{(i)} + D_2^{(i)}u^{(i)} \tag{8.7}$$

$$y^{(i)} = C_2^{(i)}x^{(i)} + D_3^{(i)}a^{(i)} + D_4^{(i)}u^{(i)}$$

where x is the state vector, a is the input vector representing the interaction from the other agents, u is the control input, z is the interface with the other agents and y is the set of measurements. The exponent $^{(i)}$ designates the ith agent.

The interaction through the power system is modeled by static feedback, $a = L$ z, where a is the vector of all $a^{(i)}$, z the vector of all $z^{(i)}$, and L a off-block-diagonal matrix that depicts the topology of the agent connections. Fig. 8.8 represents the ith and j^{th} agent interacting through the network. The uncertainty on each agent model is denoted by $\Delta^{(i)}$. It is also possible to model the uncertainty on the interface L.

The control $K(s)$ at the ith agent is restricted to be between $u^{(i)}$ and $y^{(i)}$. At each control agent, the actuator input is based on the best information available at this agent. The information $y^{(i)}$ is not restricted to be purely local information but it may not be the same at all agents. Taking into account structural constraints on measurement sets is critical in practical large system applications. The information used by an agent is tailored to its control specifications. They do not have to share a common objective, each having a specific task concurring to the same global objective.

For controller design, a variety of models, characterizing each agent, is used. Each controller accounts for the desired structure and the primary coupling effects of the neighboring agents. Information is exchanged between agents. It is important to simplify the system model to reduce the controller order and achieve robustness.

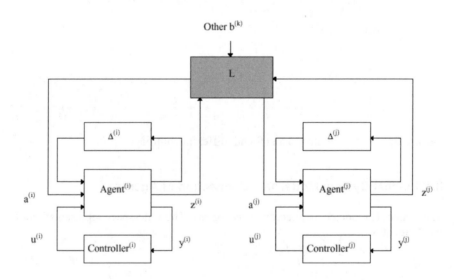

Fig. 8.8. System modeled as an interconnection of agents

8.4.2 Model Building

The control actuators are devices such as excitation systems, turbine governors, static VAr compensators, FACTS components etc. Actuators are integrated in a local or a regional subsystem defining an agent in the power system.Model reduction at each agent is an essential part of the design. Several researchers have worked on the controlled decomposition of a multi-machine power system into several subsystems.

In the simplest modeling, an area of interest consisting of one or more machines is considered in isolation as if it were connected through Thévenin impedances to a single equivalent large machine representing the rest of the system. This procedure is not satisfying in many cases and more elaborate reduction techniques have been used. Many grouping algorithms rely on coherency identification, either from transient plots or from mode shapes and participation factors.

Model building must recognize a link between model identification and output channel selection. The selection of local measurements $y^{(i)}$ is a key factor of the agent performance. Referring to Eq. (8.5), the matrix C_2 of the agent state-space representation is influenced by the measurements selected, while the matrix B_2 depends on the control agents. This suggests to design the measurement channels in terms of the combined controllability and observability effect. The model building objective is to determine the measurement matrix $C_2^{(i)}$ for the ith control agent characterized by $B_2^{(i)}$.

The model must have the lowest approximation error in a given frequency bandwidth. The model must maximize the combined impact of the measurements and of the controls in this bandwidth. The method used here looks for a SIMO model of the ith agent interacting with other agents, and it is tailored to match the agent objectives.

Several methods exist to derive such a model and the approach used is a MIMO system identification through Eigensystem Realization Algorithms. This yields a balanced system realization [9].

Once the measurements and nominal balanced models are available, an agent is able to track system changes and adapt, as necessary, the control agent.

8.4.3 Controller Building

Each agent is responsible to control its dynamics according to a local performance criterion. Each agent focuses on specific modes of the agent model. The controllers must take into account restrictions associated to time of response and robustness.

The measurements at an agent feed a controller that acts on a single input actuator. Each controller explores only a portion of the system state-space. Dynamic output feedback maintains the performance of the system model in a reference frame in which the representation is balanced. Inherent in the method, is model reduction. This limits the degree of the controller and permits to achieve a certain degree of robustness. The closed loop system behavior is approximated with reduced order dynamic compensation. The more measurements are available at an agent, the more robust and low order can be the design.

Full state feedback controllers are easy to compute for optimizing a quadratic performance index or to assign system poles; these are not practical design and they lack robustness. The method proposed here does not compromise the controller structure and remains in the time-domain [12,13]. The idea is to first define a reference state feedback controller at an agent modeled in the balanced reference frame. This is in the coordinates where an agent model is balanced. This controller is designed on a reduced order model. Since these state variables are not physically defined, and only output measurements are available, sub optimal dynamic output compensation is deduced to mimic the performance of the state feedback design. Given that each agent model is well defined, robustness is degraded mainly by the model uncertainties and not by dynamic compensation.

According to the framework, assume that control inputs $u^{(1)}$, $u^{(2)}$ ... $u^{(N)}$ must be designed for the control agents $1, 2... N$. Irrespective of the control objectives at the other agents, the control input $u^{(i)}$ is computed with a reduced order model M_i of the system which is more accurate for representing the local system behavior than the behavior elsewhere. The models M_i and M_j, and their order, are not necessarily the same. Assume that at the ith agent, the measurement set is $y^{(i)}$.

The following describes the various steps for arriving at the MMISO control in the time-domain.

- Step 1: Sequence

Select the sequence for computing each MISO controller.

- Step 2: Measurement selection

Select the measurement set, at each agent, in relation with the modes to control. From all available measurements at an agent, pick those measurements that yield an agent representation with large Hankel singular values σ and good separation according to the parameters

$$\varepsilon = (\sigma_{r+1} / \sigma_r)^{1/2} \quad \delta = (\sigma_{r+1})^{1/2} \tag{8.8}$$

Let (F,G,H,J) be the balanced realization of the transfer matrix $G(s)$. Note that (F,G,H,J) is used instead of (A,B,C,D) of Eq. (8.7) to reflect system interaction. When the poles of F are complex and with small real part, as with inter-area power system oscillations, the Hankel singular values σ_i of (F,G,H,J) are given by

$$\sigma_i = a_i \ ||H_i \ G_i|| \quad a_i = 1 / (4 \ \zeta_i \ \omega_i^2) \tag{8.9}$$

where ζ_i and ω_i are the modal damping and the modal frequency, H_i is the ith column of H, and G_i the ith row of G. Lower is the damping ζ_i, larger is a_i. The amplification of a mode can thus be controlled by the choice of measurements. The norm $||H_i \ G_i||$ is that of the entries in the residue matrix when the balancing transformation approaches the modal matrix. In a balanced realization $H_i^T H_i = G_i G_i^T$, and, when damping is small, the Hankel singular values σ_i can be approximated by γ_i defined as

$$\gamma_i = a_i \, ||H_i||^2 = a_i \, ||G_i||^2 \tag{8.10}$$

- Step 3: Time-domain performance desired

The control objective at an agent is input-output performance oriented

$$min \, J_i = min \int_0^\infty (y^{(i)T} Q_{Mi} y^{(i)} + u^{(i)T} R_{Mi} u^{(i)}) \, dt \tag{8.11}$$

where Q_{Mi} and Q_{ci} are semi-positive symmetric real matrix , and R_{Mi} is a positive symmetric real matrix. Each agent is relatively independent of the others in its choice of control objective. The matrices Q_{Mi} and R_{Mi} are selected to reflect input-output control specifications such as time constant, damping ratio, rise time of some variables etc. When $y^{(i)}$ focuses on oscillation modes, modal damping inter-pretation is immediate.

- Step 4: Computation (or identification) of a model for the ith agent

The model takes into account the controls $u^{(1)}$, $u^{(2)}$, ... $u^{(i-1)}$, $u^{(i+1)}$, $u^{(N)}$ already computed or readily available, as well as the interfaces with the agents $a^{(i)}$.

The tasks are:

- Derive a balanced system realization $M_i = (F^{(i)}, G^{(i)}, H^{(i)}, J^{(i)})$ characterizing $(u^{(i)}, y^{(i)})$;
- Reduce the system order of M_i to obtain the model $M_{ri} = (F_r^{(i)}, G_r^{(i)}, H_r^{(i)}, J_r^{(i)})$ which has the lowest truncation error possible. This model is of order n_r.

The models are not unique for a given agent and are not a complete model of the entire system.

- Step 5: Reference state feedback controller

Compute the state feedback control $K_X^{(i)}$ that optimizes the ith performance ob-jective. This assumes that all state variables of M_{ri} are available, or reconstructed.

- Step 6: Model update

Update the ith agent model so that the next MISO controller in the design se-quence takes into account feedback in M_i. This does not increase system order be-cause it involves static state feedback in the balanced reference frame.

- Step 7: Dynamic compensation

Determine the dynamic controller $C^{(i)}(s)$ which asymptotically preserves the time-domain performance achieved with state feedback for the reduced system in the balanced reference frame. This controller is a full order observer for the ith agent and it is of order n_r. This is a SISO controller when only one measurement is used, otherwise it is a MISO controller. To achieve adequate performance, the state reconstruction error must be small and the reduced model must correctly represent the system response.

- Step 8 Controller refinement

There are two alternatives to derive the final controller $C_r^{(i)}(s)$.

When there is more than one measurement at the agent, the order of $C^{(i)}(s)$, n_r, can be reduced by using the controller design freedom offered by the measurements. Reduction of the controller $C^{(i)}(s)$ order is achieved by taking into account all available measurements at the ith agent. The MISO controller order reduction achievable is the number of independent measurements $p^{(i)}$. The concept is that of a reduced order observer. As the controller order is reduced, to the limit to an algebraic controller, the controller becomes more sensitive to noise and disturbances.

This freedom can also serve to improve robustness and this is a better choice. The measurements are combined by the controller to reduce the approximation error according to the Hankel norm, to obtain a good separation $\sigma_r >> \sigma_{r+1}$ for a fixed value of r, and to shape the approximation error. The outcome is a MISO dynamic controller $C_r(s)$.

Assuming stable controllers, both $u_C(t)$ and $u_{Cr}(t)$ asymptotically tend to the reference control $u(t)$. The dynamic behavior of $C(s)$ and $C_r(s)$ depend on the measurements selection. The control is distributed among the agents. The selection of Q_{Mi} and R_{Mi} is critical for good coordination.

8.4.4 Example

The classical 3-areas WSCC system is illustrated in Fig. 8.9. The machines modeling each area are not equipped with stabilizers. The objective is to find an excitation control for machines 2 and 3 that sufficiently damps the inter-area modes, at 1.4Hz, without restricting the information to be purely local.

The linearised 9th order state model uses the well-known variables

$$x = (\omega_1, E'_{q2}, E'_{d2}, \omega_2, E'_{q3}, E'_{d3}, \omega_3, \delta_{12}, \delta_{13})^T \qquad (8.12)$$

$$u = (E_{fd2}, E_{fd3})^T$$

All system modes are shown in per unit rad/rad :

$$\Lambda = \{-0.0002 \pm 0.0001j, -0.0006 \pm 0.0230j, -0.0027 \pm 0.0346j, \qquad (8.13)$$

$$-0.0104, -0.0166, -0.0005\}$$

The most critical oscillation modes are $\{-0.0002 \pm 0.0001j\}$, $\{-.0006 \pm 0.0230j\}$ and $\{-0.0027 \pm 0.0346j\}$. The first mode, sensitive to machine damping, characterizes the overall system speed deviation. The following two modes characterize the rotor electromechanical oscillations of machines 2 and 3 with respect to machine 1.

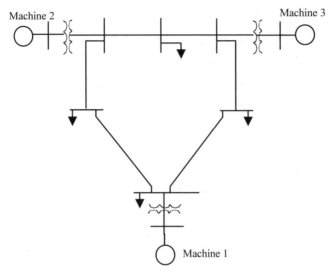

Fig. 8.9. WSCC test system

A SISO Design

The controller structure is free. Measurements selected by the agents in area 2 and area 3 are determinant for the performance. As for classical power system stabilizers, inputs are restricted to local speed, i.e. $y^{(1)}{}_2 = (\omega_2)^T$ and $y^{(2)}{}_2 = (\omega_3)^T$. According to the Hankel singular values decomposition, at both agents the model appears of order 2 and focuses on the common mode of oscillation. Inter-area dynamics is captured by expanding to a 4th order model at each agent, but the information set is not rich.

The design deals first with agent 1, then with agent 2.

First consider designing the controller for the agent $u_1 = E_{fd2}$. Assuming only the local measurement ω_2 is available, a 4th ($n_r = 4$) order SISO system model is obtained which characterizes the pair (E_{fd2}, ω_2). The set of eigenvalues Λ_{br} is shown below. Only the lowest frequency mode, the common mode, is strongly controlled. State feedback is computed in this balanced reference frame; the eigenvalues become Λ'_{br}. For the full system, and in the original reference frame, the state feedback matrix assigns the modes Λ'. The pertinent system closed loop poles are very close to those of the reduced system, validating the low order model used in system design to have a small reduction error. A 4th order SISO dynamic controller transfer function is then derived.

The closed loop poles under dynamic compensation are Λ'_{dyn}

$$\Lambda_{br} = \{-0.0002 \pm 0.0001j, \ -0.0006 \pm 0.0229j\} \tag{8.14}$$

$$\Lambda'_{br} = \{-0.0003, \ -0.0028, \ -0.0017 \pm 0.0229j\} \tag{8.15}$$

$$\Lambda' = \{-0.0004, -0.0028, -0.0016 \pm 0.0230j, -0.0027 \pm 0.0346j, \tag{8.16}$$
$$-0.0104, \ -0.0167, -0.0004\}$$

$$\Lambda'_{dyn} = \{-0.0004, -0.0028, -0.0015 \pm 0.0230j, -0.0027 \pm 0.0347j,$$
$$-0.0104, -0.0166, -0.0004, \ -0.0017 \pm 0.0219j, -0.0147, \tag{8.17}$$
$$-0.0002\}$$

In Λ'_{dyn}, the modes $\{-0.0017 \pm 0.0219j, -0.0147, -0.0002\}$ are specific to the dynamic controller.

At the second agent, the system model must take into account the control at the first agent. Assuming again only the local measurement ω_3 is available, a 4th order system model is obtained which characterizes the behavior of the pair $(E_{fd3}-\omega_3)$, with agent 1 active. The controller moves the system model poles to

$$\Lambda'_2 = \{-0.0007, -0.0027, -0.0017 \pm 0.0230j, -0.0033 \pm 0.0346j, \tag{8.18}$$
$$-0.0104, \ -0.0167, -0.0006\}$$

The mode $\{-0.0002 \pm 0.0001j\}$ is shifted even more with the second agent active. On the other hand, the interaction on the inter-area modes is weak.

A MISO Design

Controller design at agent 1 is performed for the measurement $[\omega_1 \ \delta_{12} \ \delta_{13}]^T$. The pair (E_{fd2}, δ_{12}) is used to build a new 4th order model with low error bound. Then, using additional measurements ω_1 and δ_{13} in the set, control system robustness is improved while the controller order is fixed. The idea is to iteratively shape the approximation error in a given frequency bandwidth, while maintaining a low reduction error and a good dynamics separation for the modes of interest. The procedure is accomplished by performing system balancing around a new fictitious scalar measurement z, where $z = c_1 \ \omega_1 + c_2 \ \delta_{12} + c_3 \ \delta_{13}$.

The linear combination is selected to focus the model on the truncated balanced model modes $\Lambda_{br} = \{-0.0002 \pm 0.0001j, \ -0.0006 \pm 0.0229j\}$, while reducing the approximation error at the intermediate frequencies and increasing the stability margin. The 4th order MISO controller frequency response is illustrated in Fig. 8.10 at normalized frequencies. The improvement is in stability margins. The additional measurements could be used instead to reduce the controller order, however that would generally be at the detriment of robustness.

Fig. 8.11 illustrates the control performance for an off-nominal system. Two stable controlled system responses are shown. With a 10th order H_∞ controller, the system is stable but exhibits a closed loop poorly damped oscillation. The system performance is acceptable with the MISO controller.

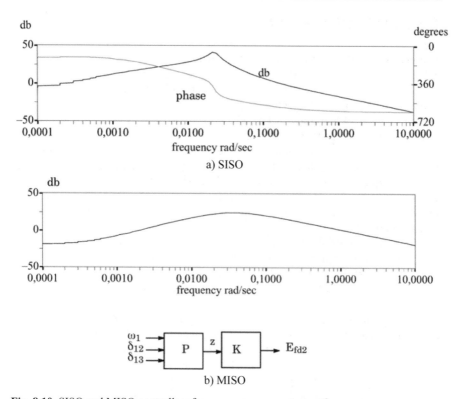

a) SISO

b) MISO

Fig. 8.10. SISO and MISO controllers frequency response at agent 2

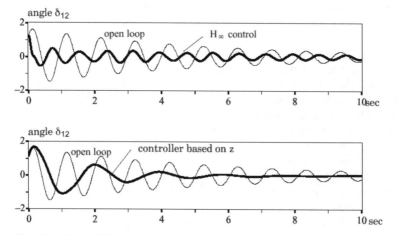

Fig. 8.11. Robust and MISO controller performances in off-nominal system

8.5 Coordination Agents

Irrespective of the MISO control at each agent presented above, the overall MMISO control needs to be coordinated to achieve a global performance objective. This objective may not be formulated as the direct sum of all the agents cooperative or competitive objectives.

The multi-level hierarchy in Fig. 8.1 helps to achieve system-wide goals. Historically, in power systems, inherent multiple time-scales have allowed a simplified form of hierarchical centralized coordination, where the actual real-time control action is performed locally and autonomously in a decentralized manner given the less-frequent centrally-computed/supplied set-points. Supervisory controls are concerned with a larger portion and broader aspects of the system behavior, dealing with slower phenomena and affording more time for decision making and set-points computation.

The slower control functions dealing with slower system dynamics such as AGC, have been implemented as on-line, closed loop and coordinated and have worked successfully. This coordination has not in general been possible for faster phenomenon, such as for power system stabilizers for damping of low frequency oscillations in power systems.

At the primary level the controllers basically regulate to their given set-points determined mainly based on off-line studies, whether local or based on synchronized phasor measurements [14]. There are many kinds of stabilizer designs to improve stability. It is impossible for fixed gain controllers to maintain the best damping performance when there is a major change in system operating conditions. Fuzzy logic and neural networks were studied for power system applications [15][16], but full on-line coordination is mainly nonexistent. Following a disturbance, the system has to have enough margin to settle to a stable equilibrium point even if it is not a desirable one and then is brought back via a reset action.

In traditional hierarchical control theories developed in the early 80's, the exact mathematical model of the controlled system is strongly relied on, and it is very difficult to design coordination strategies. Real-time model-free coordination requires new strategies to adapt the longer time performance. The coordination agents do not require the detailed structure and the parameters of the lower levels. In the hierarchical structure of Fig. 8.12, the coordinating agent is a high-level intelligent controller that attempts to modify only the reference input of the lower levels. The principle "increasing precision with decreasing intelligence" is applied. One of the main tasks associated with the coordinating agent is to design a knowledge base. An inference engine will conduct a goal-oriented search in the knowledge base according to the characteristics of system performance. The performance to step responses, or the error is reaching a target, can be used to evaluate performance. It is common usage to design the control mechanisms of agents in multi-agents systems with fuzzy controllers.

In Fig. 8.12, the control input at an agent, for example $u^{(i)}$, consists of its locally determined signal and the coordinating signal r_{ci}. For simplicity, only two regions are shown.

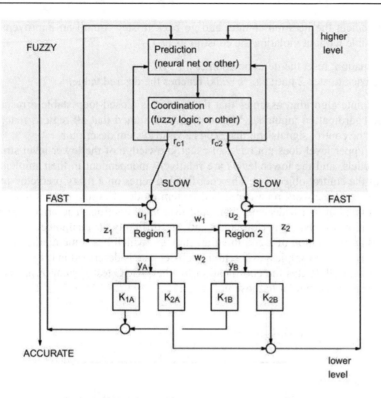

Fig. 8.12. Coordinating agent

The control is performing correctly if the output $z(t)$ is close to the model output, i.e. if the error $e(t) = z(t) - z_m(t)$ is small. Let k be the sampling instant, the coordinator thus seeks to minimize [17]

$$Jc(k) = [\, z(k+d) - zh_m(k+d\,|k)\,]^{\,T} \,\, [\, z(k+d) - zh_m(k+d\,|k)\,] \qquad (8.19)$$

where $zh_m(k+d\,|k)$ is the prediction of $z_m(k+d)$ knowing $z_m(k)$. The sequential coordination algorithm has the following steps.

Coordination of the agents
1. Given $r_{ci}(k)$, estimate $zh_m^{\,0}(k+d\,|k)$. Let the iteration counter $j = 0$

Agent 1
2. Given $zh_m^{\,j}(k+d\,|k)$, correct $r_{c1}^{\,j}(k)$
3. Given $r_{c1}^{\,j}(k)$ and $r_{c2}^{\,0}(k)$, estimate $zh_m^{\,j+1}(k+d\,|k)$
4. Increment the iteration counter and go back to step 2 until no improvement is possible without violating the constraints on r_{c1}

Agent 2: reset the iteration counter and set $r_{c1}(k)$ to the converged value
5. Given $zh_m^{\,j}(k+d\,|k)$, correct $r_{c2}^{\,j}(k)$
6. Given $r_{c2}^{\,j}(k)$ and $r_{c1}(k)$, estimate $zh_m^{\,j+1}(k+d\,|k)$

7. Increment the iteration counter and go back to step 5 until no improvement is possible without violating the constraints on r_{c2}

Coordination: reset the iteration counter
8. Go back to step 2 until $zh_m(k+d|k)$ reaches the desired value.

This simple algorithm assumes that the system is closed-loop stable, irrespective of the coordination inputs r_{ci}. Moreover, it is assumed that all regions react linearly to the control signals and that $z(t)$ is a valid system descriptor.

The upper level does not have an exact knowledge of the lower level structure and models, and the lower levels are relatively independent in their implementation of the control objectives. The coordination relies on a fuzzy predictor and expert rules. These work on a simplified system model. Coordination is in a slower time-frame but in a wider-area than the lower level controls. The price to pay is a degradation of system performance and the difficulty in guaranteeing stability. This is an illustration of a coordination strategy, which has some merits, there are alternatives, for example a different model-free one is described in [18].

Fig. 8.13 illustrates the coordination for the WSCC test system in presence of modeling errors, similar to those in Fig. 8.11.

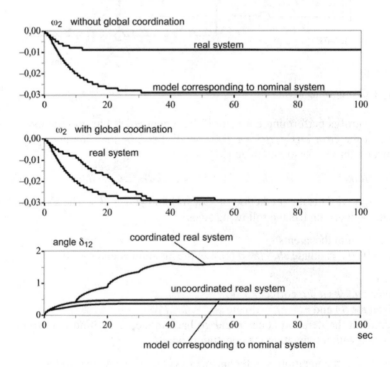

Fig. 8.13. Coordination agent in action for WSCC test system

The lower level controllers have been coordinated in the design stage based on a nominal system. The critical step of coordination is the reliable estimation of zh_m^{j+1} $(k+d \,|k)$. The event is a step change on the excitation voltage at agent 1 (the field voltage E_{fd2}).

The performance is evaluated by examining the speed ω_2 and coordination signals are issued every 10 seconds. From Fig. 8.13, under coordination, the speed ω_2 is closer to the assumed model. Since the angle δ_{12} is not a performance descriptor, under coordination its performance is worst. It is clear that the same analysis than for the damping control is necessary. Controller design at agent 1 is performed for the measurement $[\ \omega_1\ \delta_{12}\ \delta_{13}]^T$ around the pair (E_{fd2}, δ_{12}). Thus a better region 1 descriptor would be $[\ \omega_1\ \delta_{12}\ \delta_{13}]^T$.

8.5 Conclusions

Specifically, the term agent refers to a system that is autonomous in that it operates independently, that is capable of taking the initiative and that communicates and cooperates. The combination of current operating and design practices with higher-level intelligence is a promising framework to solve the control problems in power systems. The multi-agents system approach offers a regionally decentralized control model with global coordination.

In this framework, the upper level of hierarchical control should concentrate its attention to the global coordination and optimization and not excessively interfere the controls in the lower levels. The tasks, to achieve a global goal, are distributed among agents and each agent is to perform its special task. The long-term objective is fully coordinated and require closed-loop high-bandwidth power system control capabilities. It is reasoned that the dynamic information infrastructure will draw upon a hierarchical monitor network that provides integrated and fairly comprehensive measurements of wide area dynamics. The initial application of intelligent agents for real-time control of the power grid may focus on Control Area functions related to slower Energy Management System functions.

In a large power system, the monitoring of local devices or even regional performance remains a challenge. Reliable data capture, for example, implies sub-networks of continuously recording monitors plus intelligent information tools that can cope with the very high data volumes. What is introduced in this chapter is just preliminary research. To accomplish the system, there is still a long way to go. There are important questions remaining, such as: What is the best strategy and structure to coordinate multi-agents systems? What is the knowledge necessary for coordination? How can we analyze and guarantee system stability? How can we guarantee that malfunctions in individual agents will not move control away from the objectives?

References

[1]   Trudel G, Bernard S, Scott G (1999) Hydro-Québec's defense plan against extreme contingencies. IEEE Trans. on Power Systems, vol 14, pp 958–966.

[2]   Taylor CW, Lefebvre S (1991) HVDC controls for system dynamic performance. IEEE Trans. on Power Systems, vol 6, pp 743-752.

[3]   Kundur P (1994) Power system stability and control. McGraw-Hill.

[4]   CIGRE Task Force 38.01.07 (1996) Control of power system oscillations. CIGRE Technical Brochure.

[5]   IntelliTEAM ® Overview of Operations. EnergyLine Systems, Alameda, CA, www.energyline.com

[6]   Huang JA, Valette A, Beaudoin M, Morison K, Moshref A, Provencher M, Sun J (2002) An intelligent system for advanced dynamic security assessment. International Conference on Power System Technology, vol 1, pp 220 –224.

[7]   Wildberger AM (1997) Complex adaptive systems : concepts and power industry applications. IEEE Trans. on Control Systems, December, pp 77-88

[8]   Hauer JF (1991) Application of Prony analysis to the determination of modal content and equivalent models for measured power system response. IEEE Trans. on Power Systems, vol 6, pp 1062-1068

[9]   Sanchez-Gasca JJ, Chow JH (1999) Performance comparison of three identification methods for the analysis of electromechanical oscillations. IEEE Trans. on Power Systems, vol 14, pp 995-1002

[10]   Kamwa I, Trudel G, Gerin-Lajoie L (1996) Low-Order black-box models for control system design in large power systems. IEEE Trans. on Power Systems, vol 11, pp 303-311

[11]   Yusof SB, Rogers GJ, Alden RTH (1993) Slow coherency based network partitioning including load buses. IEEE Trans. on Power Systems, vol 8, pp 1375-1382

[12]   Lefebvre S (2000) Dynamic output compensation between selected channels in power systems. Part 1: The framework. Electrical Power and Energy Systems, vol 22, pp 155-163

[13]   Lefebvre S (2000) Dynamic output compensation between selected channels in power Systems. Part 2: Applications. Electrical Power and Energy Systems, vol 22, pp 165-178

[14]   Kamwa I, Grondin R, Hebert Y (2001) Wide-area measurement based stabilizing control of large power systems - A decentralized/hierarchical approach. IEEE Trans. on Power Systems, vol 16, no 1

[15]   Park YM, Moon UC, Lee KY (1996) A self-organizing power system stabilizer using fuzzy auto-regressive moving average (FARMA) model. IEEE Trans. on Energy Conversion, vol 11, pp 442-448

[16]   Park YM, Choi MS, Lee KY (1996) A neural network-based power system stabilizer using power flow characteristics. IEEE Trans. on Energy Conversion, vol 11, pp 435-441

[17]   Cui X, Shin KG (1991) Intelligent coordination of multiple systems with neural networks. IEEE Trans. Man and Cybernetics, vol 21, pp 1488-1497

[18]   Ning W, Shuqing W (1997) Neuro-intelligent coordination control for a unit power plant. IEEE International Conference on Intelligent Processing Systems, vol 1, pp 750-753

9 Development of a Coordinating Autonomous FACTS Control System

Christian Becker

Institute of Electric Power Systems, University of Dortmund, Germany

Novel power-electronic FACTS devices (FACTS - Flexible AC Transmission Systems) can be used for fast, stepless adjustable power flow control and voltage control as well as for the improvement of stability in electrical power systems. Therefore, they are attractive especially for grid companies in liberalized energy markets. However, the uncoordinated use of FACTS devices involves some negative effects and interactions with other devices, which leads to an endangerment of the steady-state and dynamical system security. Up to now, these problems have led to a very restricted acceptance of the grid companies for FACTS devices. The present chapter gives an analysis of the necessary coordinating control measures in FACTS and expresses them as abstract coordinating rules. Based on that, an autonomous control system for electrical power systems with embedded FACTS devices is developed that realizes the necessary preventive coordination. With methods of computational intelligence the system automatically generates specific coordinating measures from the abstract coordinating rules for every operating condition of the power system without human intervention or control. That way an optimal utilization of the technical advantages of FACTS devices as well as the steady-state and dynamical system security is guaranteed. Interactions between the autonomous system and other existing controllers in electrical power systems are taken into consideration during the development process so that the autonomous system can completely be integrated into an existing conventional network's control system.

9.1 Introduction

The worldwide trend towards the liberalization requires electric power systems with a defined reliability under uncertain future operating conditions and scenarios. It also requires an increase of the exploitation and flexibility of existing power systems. Independent of this, several technical restrictions, as e.g. congestions, loop flows, inadequate fault tolerances, and stability limits, affect an efficient use of today's power systems. Against this background the application of new power electronic devices in electrical power systems becomes increasingly attractive. These devices are familiar under the name "FACTS devices" (FACTS - Flexible AC Transmission Systems) and enable an expanded controllability of the system [1].

Features, structure, and control of FACTS devices are described more closely in Section 9.2. At present, FACTS devices are in many cases still in the development phase. However, first prototypes of FACTS devices are already in use for system studies [2]. They enable interventions into the system in short-term range and are implemented for fast power flow control, for voltage control as well as for the improvement of stability [3]. Nevertheless the use of FACTS devices in power systems causes the following problems that have to be solved:

The response time of FACTS devices is in the range of some ten milliseconds. In case of critical events within the power system, e.g. faults or overloadings, FACTS devices react immediately to these events due to their short response time. As the FACTS devices are not adapted to the new situation after such a critical event, this can lead to an endangerment of the steady-state and dynamical system security [4].

As a consequence, the application of FACTS devices requires both a fast coordination of their controllers among one another and with power plants, loads, and conventional controlling devices within the power system. This coordination must guarantee the steady-state and dynamical system security in the case of critical events and has to be automatic, quick, intelligent, and preventive.

In Section 9.3 the necessary coordinating measures in any power system with embedded FACTS devices are derived from a system theoretical point of view using a basic network configuration and are explained in detail. They can be expressed qualitatively in the form of generic rules prescribing general actions. They are not formulated device-specific [5].

As a solution a coordinating autonomous control system is developed basing on the coordinating generic rules. Section 9.4 shortly summarizes the theory of autonomous systems, which bases on [6] and which is presented in detail in chapter 1. Autonomous systems generally represent an abstract information-technological framework, which is specified in this paper for the coordinated control of FACTS devices.

Section 9.5 describes the implementation of the autonomous control system. Two of the necessary coordination measures basing on the above mentioned generic rules are carried out by the autonomous system with the help of fuzzy modules. They intervene adaptively into the local FACTS-controllers. Different controllers can also be activated or deactivated by this fuzzy adaptation. Additionally, outer control loops perform coordination according to another rule by changing the setpoint values after the occurrence of critical events. The specific coordinating measures of the fuzzy modules and outer control loops are derived from the generic coordinating rules as well as from the actual operating state and the topology of the electric power system.

Due to the continuous changes of the operating states and the topology during the daily operation through varying loads, generations and switching operations, the specific coordinating control measures and damping controller parameters must be followed up automatically to these changes. Only under this condition the fuzzy modules, additional outer control loops and damping controllers are able to react adequately on critical events in the changed system. This guarantees a dynamical and stationary secure behavior of the whole system as well as sufficient system stability. To ensure a quick reaction of the autonomous system, the derivation of the specific coordinating measures has to be performed, before a critical event occurs. Hence, topology-changes of the network have to be analyzed continuously. This continuous adaptation of the specific coordinating measures to changing topologies is called "preventive coordination" [4] being performed by the autonomous control system. Section 9.6 finally presents examples for the application of the autonomous system, which are verified by dynamic simulations of an example power system.

In comparison to the state of the art the method proposed within this contribution is a basic new approach. In the past, controllers for FACTS devices with different tasks are globally parameterized, but only for the normal operation and a fixed network topology [7, 8]. Therefore, the reaction on critical events is insufficient and maloperations and an endangerment of the system security can arise as soon as the power system is transferred into a different operating state.

This contribution bases on the research work, which has been published in [4]. The chapter is not intended to deliver a complete and detailed description of how to implement the derived and explained methods. It shall rather reveal how the theory and idea of autonomous systems can be applied for the development of an intelligent coordinating control system for FACTS. Details of the methods, which are important for the implementation of the complete autonomous system, can be found in [4].

9.2 Flexible AC Transmission Systems – FACTS

9.2.1 Features

FACTS devices offer an expanded controllability of power systems through their electrical characteristics of shunt- and/or series-compensation and phase shift control, which is realized through fast current-, voltage-, and impedance-control. Essential elements of FACTS devices are power electronic components, which enable a very short response time in the range of a few ten milliseconds [9].

FACTS devices are employed for load flow control, voltage control and the improvement of stability [3]. The UPFC (Unified Power Flow Controller) is a universally applicable FACTS device, because it combines the characteristics of most of the other FACTS devices [10]. Therefore, only the UPFC is regarded in this chapter for the development of an autonomous control system for FACTS.

9.2.2 Structure

The electrical characteristics shunt-/series compensation and phase shift control are realized through the application of modern power electronics together with a shunt- and a booster-transformer [9].

The structure of FACTS devices can be schematically illustrated by quadripoles [4]. Fig. 9.1 shows the structure of the UPFC as a serial-parallel FACTS device. The two voltage-sourced switching converters (VSC_p and VSC_s) are connected to the shunt- or the booster-transformer respectively and are linked with a d.c.-component, which is represented by a capacitor. In serial FACTS devices there is only one transformer in the series branch, in parallel ones accordingly only one in the shunt branch.

9.2.3 Modeling and Control

Through its local operating point controllers a UPFC is able to control active- and reactive-power flow over the transmission line, to which it is installed in se-ries, and the nodal voltage at one of its connection-points i or j (see Fig. 9.2). Usu-ally these three controllers are implemented as PI controllers. They receive set-point values for the active- and reactive-power flows $P_{ij,ref}$, $Q_{ij,ref}$, and nodal voltage absolute values at the connection points of the UPFC $V_{i,ref}$ or $V_{j,ref}$ respec-tively as well as the corresponding feedback variables.

The outputs of the three controllers are the real and the imaginary part of the longitudinal voltage $\overline{V}_l$ to be impressed, and the absolute value of the shunt cur-rent I_q which has to be injected.

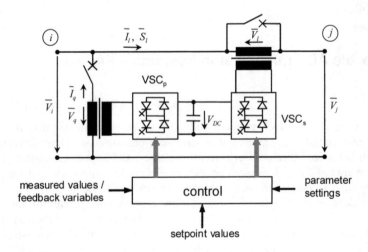

Fig. 9.1. Schematic Structure of a UPFC

The active power flow controller determines the imaginary part whereas the reactive power flow controller determines the real part of the longitudinal voltage $\overline{V}_l$. This is due to the strong coupling between active power and voltage angle and reactive power and absolute value of the voltage. From these outputs and further local measured values of the UPFC, such as the DC voltage V_{DC} and the voltage angle of $\overline{V}_i$, the converter control determines the trigger angles for the GTO-Thyristors of the two VSCs and generates impulses for their activation. Furthermore, it keeps the limits of the manipulated variables. In this way the shunt and the booster transformer generate the necessary voltage or current respectively in order to keep the setpoint values.

As shown in Fig. 9.2 the dynamic behavior of a UPFC can be modeled by a current source injecting the shunt current $\overline{I}_q$ and a voltage source impressing the longitudinal voltage $\overline{V}_l$. The dynamics of the two VSC are modeled by first order time delay elements (PT_1-Elements) with a time constant in the range between 15 and 30 ms [11].

In the model, the outputs of the operating point controllers are therefore not used to compute trigger-angles for the VSCs. Instead of this their output can be directly used by the converter control model for the calculation of $\overline{V}_l$ and $\overline{I}_q$ which have to be applied within the UPFC model. Furthermore a controller for improving the small-signal-stability of the system (damping controller) can be present. It is usually implemented to work in parallel to the operating point controllers (see Fig. 9.2), which means that the outputs of the damping-controller are added to the outputs of the operating point controllers. In [4] an output feedback controller with adaptable parameters has been examined and shown excellent improvements of the system's small-signal-stability. This controller is linear and it is usually parameterized for a power system model being linearized around an operating point. Constant parameter settings can usually only guarantee good control performance for the system operating around this point and not within the whole range of states in which it can operate. Hence, the damping controller parameters are to be adapted to changes of the system's state. This will be further regarded in Section 9.3. The outputs being fed back by the controller are the deviations from the setpoint values of active-power (ΔP_{ij}), reactive-power (ΔQ_{ij}), nodal voltage (ΔV_i) and the corresponding serial current (ΔI_l). The controller function is defined in Eq. 9.1. Its input and output vectors are defined in Eq. 9.2 and Eq. 9.3.

$$\Delta u = -F\,\Delta y \tag{9.1}$$

$$\Delta y = \left(\Delta V_i \;\; \Delta Q_{ij} \;\; \Delta P_{ij} \;\; \Delta I_l\right)^T \tag{9.2}$$

$$\Delta u = \left(\Delta u_{V,D} \;\; \Delta u_{Q,D} \;\; \Delta u_{P,D}\right)^T \tag{9.3}$$

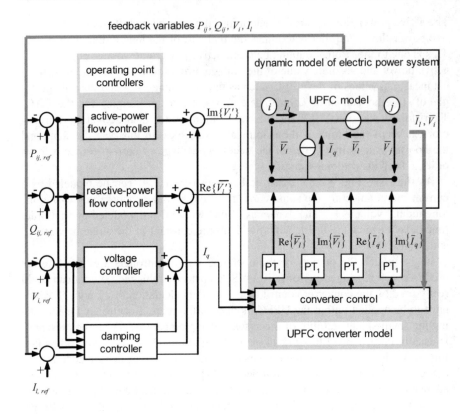

Fig. 9.2. UPFC modeling and control

9.3 Need of Coordination

An extensive system theoretical analysis of power systems containing FACTS devices has been carried out in [4]. This reveals that the use of FACTS devices in electric power systems in terms of a maximal exploitation of their technical and economical advantages requires a coordination of their controllers among one another and with power plants, loads and conventional controlling devices. Otherwise the dynamic behavior of the power system gets worse and an endangering of the steady-state and the dynamical system security is to be expected. Coordination is essential for the steady-state as well as for the dynamical operation.

Coordination for the steady-state operation can e.g. be performed using optimal power flow techniques [12]. Concerning the dynamical operation, an adaptation of the control operations by FACTS devices to changing operating situations or critical events in the power system has to be performed.

In the following, concrete (critical) events, which require coordinating control measures to be applied to the embedded FACTS devices, are listed:

- overloading of electrical devices,
- failure of electrical devices,
- short circuits in transmission elements,
- change of the system's state.

Necessary coordinating control measures have to be applied in short term range after the occurrence of one of the above-mentioned (critical) events. They can be formulated in a knowledge-based form as so-called generic rules [5]. Before they will be listed and explained, the definition of the terms "control path" and "parallel path", which concern the network topology, has to be given (see Table 9.1). For illustration, the topology of a simple example power system including one UPFC is shown in Fig. 9.3.

The existence of a parallel path is an essential necessity for a power flow controlling FACTS device. Controlling the power flow over its control path a FACTS device shifts the power flow from its control path to parallel paths and vice versa.

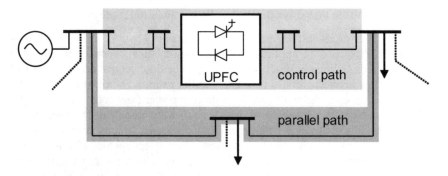

Fig. 9.3. Simple example power system used for definition of control and parallel paths

Table 9.1. Definition of terms

Term	Definition
control path	transmission path in which a power flow controlling device (e.g. UPFC) is implemented and which only has junctions at its end-nodes
parallel path	transmission path which starts and ends at the same nodes as a control path and in which no power flow controlling device is implemented

The following four coordinating control measures have been found out during the above mentioned system theoretical analysis and will be treated within this chapter:

1. **IF** a device on a parallel path of a FACTS device is overloaded,
   THEN modify the *P*-setpoint-values of the FACTS-device

A power flow controlling FACTS device can directly influence the active and reactive power flow over its control path. This leads to the above-mentioned shift of the power flow from the control path to parallel paths or vice versa. Consequently, power flows over parallel paths can be specifically influenced by changing the setpoint values for the active- and reactive-power flow of the control path. In this way overloadings of devices on parallel paths can be suppressed by changing the setpoint values of a power-flow controlling FACTS device. The control path takes over the surplus of power flow which otherwise leads to the overloading of the device(s) on a parallel path.

This rule recommends to modify only the *P*-setpoint-values of FACTS devices to suppress overloadings because these are mainly caused by active power flows. The reactive power-flow controlling functions of a FACTS device can then be used for voltage control.

2. **IF** there is a failure of a device on a parallel path **AND** no further parallel path exists for a FACTS device
   THEN deactivate the power flow controllers of the FACTS device

The existence of at least one parallel path to a control path is an important condition for the reasonable application of the power flow control function of a FACTS device. As already described above, power flow control causes a shift of the power flow between control path and parallel path(s). Hence, if a failure of a device causes an opening of all parallel paths, the power flow control of a FACTS device is hindered. The consequence would be that the outputs of the FACTS device's power flow controllers would run into their limits, which may cause strong system oscillations. This is called "false controlling effect", which means that the power flow controllers try to meet the given setpoint values, but they cannot reach them because power flow can not be shifted to parallel paths. By quickly deactivating the power flow controllers after such a failure the false controlling effect can effectively be prevented.

3. **IF** a short circuit happens on a control path or on a parallel path of a FACTS device,
   THEN slow down the operating point controllers of the FACTS device

This coordinating measure prevents excessive power oscillations after a short circuit followed by automatic reclosing. The reason for this is that the power flows change strongly during the short circuit. Mainly, a high reactive current flows over every line into the direction of the short circuit location. Because of the short response time of the FACTS devices the power flow controllers respond immediately to the short circuit and try to meet the preset setpoint values. Also the voltage controller tries to fix the setpoint-voltage. Hence, the outputs of the operating point controllers will strongly increase within a short period of time and reach their limits even before the fault is clarified and the automatic reclosing is started. When the fault is removed after an automatic reclosing these large values of the manipulated variables of the operating controllers lead to strong oscillations. This is another kind of false controlling effect and has to be suppressed by suitable measures. Through slowing down the power flow controllers and the voltage controller during the short circuit and the automatic reclosing (decreasing of the PI controller parameters) this false controlling effect can be prevented.

4. **IF** a change of the dynamical state of the entire system happens,
   THEN adapt the parameters of the FACTS device's damping controller

From a control systems theory point of view, an electric power system behaves like a non-linear, time-variant controlled system. Due to permanent changes of power generation, loads, and the networks topology, the dynamical state of the system varies strongly and continuously. FACTS devices which are equipped with simple damping controllers such as the above mentioned linear output feedback controller are installed to improve the small-signal-stability of a power system during all its operating conditions. Therefore they have to be continuously adapted to the changes of the dynamical state, which happen to the entire system.

The correct application of these four coordinating measures to FACTS devices and their control enables the network operators to exploit the advantages being offered by FACTS and their steady-state and dynamical secure operation. An autonomous control system shall be designed to realize them automatically.

9.4 Theory of Autonomous Control Systems

Chapter 1 of this book comprehensively explains the essential ideas of the autonomous systems theory. Basing on that this paragraph therefore summarizes very briefly the most important aspects of the theory that are used for the development of the autonomous control system for coordinating control of FACTS.

An autonomous control system is a control unit of a technical system. Generally its architecture can be subdivided into several intelligent autonomous components communicating with each other. The autonomous components themselves consist of different authorities called 'management', 'coordination', and 'execution'. Depending on the control level on which an intelligent autonomous component is placed, one of the three authorities dominates compared to the other two authorities. In order to specify the components on each control level every necessary local controller of the process must be determined concerning its structure. An autonomous component can be a control station, a process computer or a simple controller.

The control level on which a component is placed also determines the main functionality of the component. In [6] a general concept of autonomous control systems is presented consisting of several intelligent controllers working in parallel and independently. This concept can be adapted to a more flexible architecture of autonomous control systems in electrical power systems [13].

According to the hierarchical model of an instrumentation system for complex technical processes, e.g. electric power systems, the different control levels are called:

- network control level,
- substation control level,
- bay control level.

This leads to the structure of the autonomous system shown in Fig. 9.4, where the actuators can be represented by FACTS devices e.g.. In the following the essential characteristics of autonomous components on the different control levels are briefly described.

Bay Control Level: The physical coupling of the autonomous components on the bay control level is realized by sensors and actuators. The main task at the bay control level is 'execution', i.e. in this context mainly the application of control and adaptation algorithms.

Substation Control Level: Autonomous components on the substation control level mainly act as coordinators. They determine and plan the functionality of other components and delegate distinct special tasks.

Network Control Level: On the network control level autonomous components are working with information being generated from a model of the whole process, which can be implemented on this control level. The most important task of these components is the decomposition of global aims being generated here or prescribed by a human operator through the human-machine-interface.

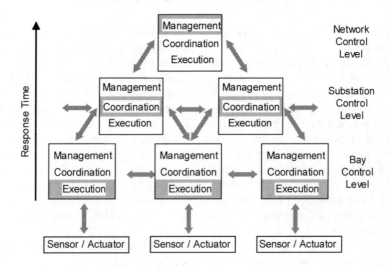

Fig. 9.4. Autonomous Control System

This theoretical concept of autonomous systems is independent of the kind of physical process. It leads to the following features and advantages of autonomous systems:

- They can perform self-learning, self-organization, and can plan and optimize control actions.
- Artificial intelligence can be arranged decentrally. This enables quick autonomous intelligent actions.
- Autonomous systems can adapt themselves automatically to changes of the technical process in structure and parameter.
- The operation of the process can be performed without the need of human intervention or monitoring.

Consequently, faults can be localized quicker, the exploitation of the process will be increased, the adaptability of the whole system will grow, and reaction times as well as data flows between decentral units and the control center will be reduced.

9.5 Synthesis of the Autonomous Control System for FACTS

The four coordinating generic rules, which have been explained in Section 9.2, are the elementary tasks, which have to be fulfilled by the autonomous control system, whose development is described in the present section.

The first three rules mainly concern setpoint values for the operating point controllers and the operating controller's parameters, whereas the fourth rule is related to the damping controller. For this reason the last rule will be treated separately from the first three rules. The development of the autonomous system is performed successively starting at the lowest control level, i.e. the bay control level. Some elementary autonomous components are chosen and designed to be acting on this control level. After that, additional autonomous components on the other control levels are added. They provide the components on the bay control level with necessary specific information, which is generated automatically in dependence on the actual network's topology or the system's dynamical state.

9.5.1 Bay Control Level

Generic Rules 1, 2 and 3

The coordinating measure given by the first generic rule requires a modification of the P-setpoint value of the UPFC in order to prevent overloadings on lines on its parallel paths. A simple but effective autonomous component performing this can be an integral-action controller forming an outer control loop. The actual active-power flows over all lines on parallel paths have to be observed by the autonomous component. As the degree of freedom for influencing power flows over parallel paths of one FACTS device equals one, a UPFC can at the same time specifically prevent only one overloaded line. If one or several overloadings are detected, the line with the biggest overloading is chosen. The actual deviation from the maximum allowed active power flow (P-P_{max}), which has a positive value in case of an overloading, is taken as the input of the integral-action controller. This way it adjusts the setpoint of the active-power flow controller of the UPFC until the active-power flow of the overloaded line is reduced to its maximum allowed value P_{max}. This is the basic idea of how the first generic rule is realized on the bay control level. It guarantees the steady-state security of the power system.

When using this method in practice several additional measures have to be implemented. This comprises e.g. the detection if the reason of an overloading has been disappeared after the overloading has been removed by the integral-action controller. In this case the setpoint adjusting by the integral-action controller has to be reset. Another important thing is the detection if an overloading is permanent or only temporary. Temporary overloadings can appear in case that the active power flow over a line oscillates around a value, which is directly below the maximum capacity. Those temporary overloadings are usually uncritical because they do not cause thermal problems. Hence they do not have to be treated by the autonomous control system. Additionally, it has to be respected that not all overloadings of lines on parallel paths can be removed by the P-setpoint adjusting. It strongly depends on the impact of a FACTS device on the power flow of parallel paths, which can be high or very low. In case the impact is very low, usually a very big change of the P-setpoint is required for removing the overloading. As the UPFC has only limited control power, the setpoint adjusting will probably not be successful when trying to remove the overloading. These and further specific aspects are very important for the implementation of the method [4].

The second generic rule requires a deactivation of the power flow controllers in case of failures of distinct devices on parallel paths. The deactivation of the controllers shall be performed by quickly setting the controller parameters of the active and reactive power flow controller to zero. Adaptive control is chosen to be suitable for this. Since fuzzy adaptation provides a transparent knowledge based implementation of adaptation rules, a fuzzy module is chosen to be the autonomous component on the bay control level performing this task (fuzzy module 1).

In addition, such a fuzzy adaptation produces soft transitions between the activation and deactivation of the controllers. The knowledge bases are derived from the generic rule 2. This is performed by autonomous components on higher control levels and will be described in Sect. 9.4.2. The input quantities of the fuzzy controller must be measured values of lines on parallel paths. From these input quantities the fuzzy controller must be able to clearly recognize failures of relevant transmission elements. Measurements of the currents or complex power flows over the concerning transmission elements can be taken as input quantities. Membership functions for the input quantities have to be chosen once and remain valid for all operating cases [4].

The implementation of the third generic rule on the bay control level is also done by a fuzzy controller performing an adaptation of the operating controller parameters (fuzzy module 2). It decreases the operating point controller's parameters in cases of short circuits on lines on the control path or on parallel paths so that the controllers are slowed down strongly, as it is required according to generic rule 3.

Short circuits (faults) must be reliably recognized by the input quantities of the fuzzy controller. Hence, the currents over those lines can be taken as input quantities for the fuzzy controller e.g. Also here the membership functions have to be chosen only once.

Fig. 9.5 shows the operating point controllers of a UPFC which are extended by the explained additional controllers as autonomous components on the bay control level. They realize the basic measures, which are required by the first three generic rules.

Generic Rule 4

As already described in Sect. 9.2.3 the damping controller is designed as an output feedback controller whose feedback matrix F has to be adaptable to changing conditions of the power system according to generic rule 4.

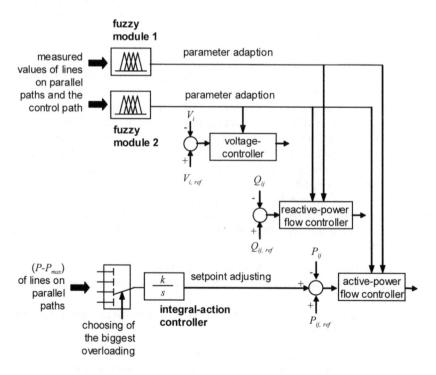

Fig. 9.5. Operating point controllers of a UPFC with autonomous components on the bay control level

The bay control level does not contain any specific autonomous components for the adaptation of the damping controller since information about the whole system's state can only be provided from the entire power system's point of view, i.e. from the network control level.

9.5.2 Substation and Network Control Level

Autonomous components on the substation and the network control level have to generate specific additional information for the autonomous components on the bay control level (fuzzy modules, integral-action controller and damping controller). This must also be based on the four generic rules. Similar to Section 9.4.1, generic rule 4 will be treated separately.

Generic Rules 1, 2 and 3

The generic rules strongly depend from the network's topology. They use the terms "control path" and "parallel path" as they have been defined above. For this reason, autonomous components on the network control level have at first to analyze automatically the network's topology. This is done recursively with the well-known backtracking technique [14]. The result is an assignment of all parallel paths to each control path. For large and complex networks these calculations can take long computation time because theoretically a large number of parallel paths may exist. However, since the impact on parallel paths that are far away from the control path may be very small, the user can define a reasonable area of impact for each FACTS device, in which it has sufficient impact on its parallel paths. These areas should be chosen such that the influence of the power flow over lines within the areas can be performed with a realistic amount of control power. The analysis of the network's topology for finding control and parallel paths can then be limited to these areas of impact.

With the result of the topology analysis the three generic rules can be brought to a set of concrete coordinating rules, which are valid for the actual network topology. To illustrate this, one example of a concrete rule for each generic rule shall be given:

1. **IF** line 11-19 is overloaded **THEN** modify the *P*-setpoint-values of UPFC 2
2. **IF** there is a failure of line 17-18 **THEN** deactivate the power flow controllers of UPFC 1
3. **IF** a short circuit happens on line 11-19 **THEN** slow down the operating point controllers of UPFC 2

This is how the rules may look like for an example real power system containing UPFCs. The complete sets of concrete coordinating rules may contain a large number of rules.

For the generic rules 2 and 3 the concrete rules are then translated by autonomous components into fuzzy rule bases for the fuzzy modules 1 and 2 on the bay control level for each FACTS device. The rule bases are downloaded into the fuzzy modules 1 and 2.

Concerning generic rule 1 the result of the topology analysis is used by a further autonomous component to compute the impact of the FACTS devices on lines on parallel paths. It computes the GSDF (generation shift distribution factors, [15]) in order to quantify the impacts of FACTS devices on all lines of the parallel paths. Only if the impact of a FACTS device on a line is big enough, it is sensible to include this line into the autonomous control in terms of preventing overloadings. If more than one FACTS device has a certain impact on a line, the FACTS device with the biggest impact on that line is determined to remove a possibly occurring overloading. This way the GSDF determine the lines, which have to be monitored by which FACTS device with regard to overloadings. They also determine the parameters k of the integral-action controllers. This mainly concerns the sign of the control action, which means if the P-setpoint has to be increased or decreased to remove a specific overloading of a transmission element.

In this way it can be guaranteed that the integral-action controllers perform their control actions to remove overloadings with the correct direction and the necessary intensity.

Fig. 9.6 finally shows the autonomous components which are necessary on the substation and the network control level in order to generate specific information for the fuzzy modules and the integral-action controllers as autonomous components on the bay control level.

Generic Rule 4

Autonomous components on the network and substation control level have to determine the damping controller parameters, i.e. the elements of the output feedback matrices F of each FACTS device being fitted with such a damping controller. As already done in the previous sections of this chapter, the ideas and concepts of how this is performed shall be revealed instead of presenting all details about their implementation.

As mentioned above, loads, generations and the network's topology determine the dynamical state of the power system as a controlled system, its input variables and its equivalent transmission function. The non-linear system equations for a current operating point can be linearized around this operating point, such that a set of linear coupled differential equations is received. Hence, the power system can be described as a first-order state space model, which is valid in a certain environment around the chosen operating point.

The computation of the eigenvalues of the system matrix A gives information about its oscillatory characteristics, e.g. critical modes. Critical oscillation modes are modes with a small or even negative damping ratio. Furthermore, the eigenvalues have to be computed by an autonomous component in order to determine the modal transformation of the system. This is also done on the network control level. Regarding the input matrix B_m of the modal transformed system it can easily be analyzed, which FACTS damping controller has got a strong influence on which of the critical oscillation modes of the system.

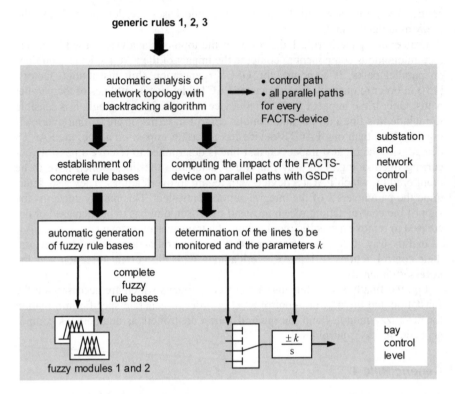

Fig. 9.6. Autonomous components on the substation and network control level for generic rules 1, 2 and 3

The autonomous components assign to the critical mode with the lowest damping ratio one FACTS device, whose damping controller has the biggest influence on that mode. The remaining FACTS devices are then one by one assigned to other critical modes with higher damping ratios.

Using this selection and some further information, like the damping sensitivity factors (DSF) [16], a cost function is formulated which expresses the effectiveness of a chosen parameter set for the FACTS damping controllers, concerning the resulting damping ratios of the critical modes in the closed-loop operation. This cost function is then minimized using the well known Simulated Annealing algorithm as a numerical optimization technique [17] in order to determine the optimal output feedback control matrices F_i for each existing FACTS damping controller and the present system's state. Fig. 9.7 illustrates the whole described procedure being performed by autonomous components on the network control level in order to compute optimal FACTS damping controller parameters after a change of the dynamical state of the entire system.

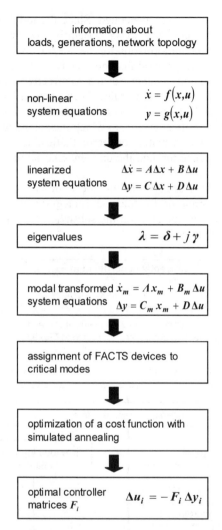

Fig. 9.7. Procedures being performed by autonomous components on the network control level for the automatic adaptation of FACTS damping controller parameters after changes of the dynamical system state (index i denotes the i-th FACTS device, if several FACTS devices are installed in the power system)

9.5.3 Preventive Coordination

As already mentioned in Sect. 9.1, the specific information for fuzzy and integral-action controllers (fuzzy rule bases etc.), which represent the coordinating measures in the case of overloadings, faults, and failures, is only valid for the network topology for which they have been generated. Since the topology of the system changes in the daily operation of the power system by switching operations, the fuzzy rule bases and additional information for the integral controllers must be followed up automatically to these modifications. Only under this condition it is guaranteed that the autonomous system can react correctly to critical events according to the above-mentioned generic rules. Such planned changes of the network's topology are named in Fig. 9.8 with "intended topological changes". In addition, the occurrence of critical events, to which the autonomous system reacts by means of fuzzy parameter adaptation or setpoint adjusting, itself may lead to a changed topology, for instance through the unintentional failure of a transmission line.

For both cases of topology changes the previously described procedures for the generation of specific information for fuzzy and integral-action controllers being performed by autonomous components on the substation and the network control level have to be activated automatically.

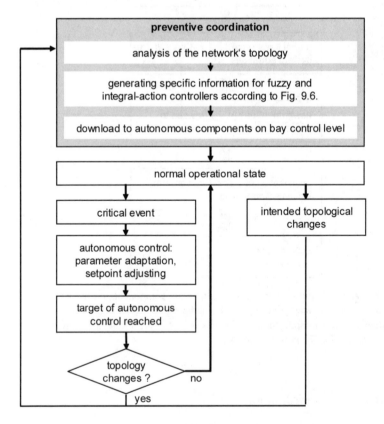

Fig. 9.8. Procedure for preventive coordination of FACTS devices

Hence, new specific information is generated for the autonomous components on the bay control level. This is called "preventive coordination". The term "normal operational state" means that at the moment no certain coordinating action of the autonomous system is required so that only the operating point controllers are in normal operation. "Target of autonomous control reached" indicates that an overloading has been successfully removed or that operating point controllers have successfully been slowed down or deactivated in order to prevent false controlling effects respectively.

9.6 Verification

In the following, two simulation examples are shown in order to illustrate the performance of the autonomous control system. For the investigations the example network according to Fig. 9.9 has been analyzed by means of dynamic simulations using a MATLAB®/SIMULINK® simulation environment. The fuzzy rule bases and global information for integral controllers were generated with the described autonomous system, which has also been implemented into the simulation environment.

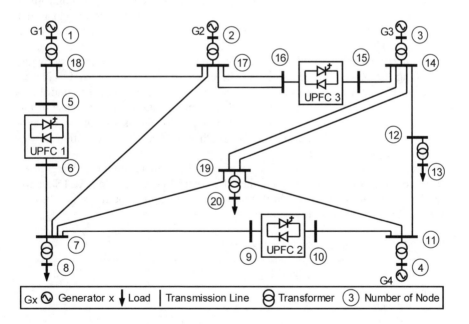

Fig. 9.9. Topology of the test system

A failure of a transmission line and the scenario of a line overloading caused by a rapid increase of a load is simulated. In this way the effectiveness of the coordinating measures through generic rules 1 and 2 is shown. The per-unit quantities of the used example power system are: $S_b = 1250$ MVA and $U_b = 400$ kV. Simulation examples showing the effectiveness of the realization of the generic rules 3 and 4 by the autonomous control system can be found in [4].

The example system is derived from the extra-high voltage level of large power system, which has been reduced to the essential transmission elements, generations and loads.

9.6.1 Failure of a Transmission Line

A failure of line $L_{17\text{-}18}$ is assumed. It occurs at $t = 0.1$ s with a duration of 4.9 s. Line $L_{17\text{-}18}$ has before been correctly identified as a part of a parallel path of UPFC 1 by the topology analysis. The results of the automatic topology analysis are listed in Table 9.2. As the control path of UPFC 1 has only got one parallel path, this means that there is no parallel path existing after the failure of this transmission line. Without autonomous control the controllers try to keep the setpoint value for active and reactive power flow over UPFC 1 and produce a large value for V_l up to its limit of 0.15 pu (see Fig. 9.10). This is due to the false controlling effect. However, the setpoint values cannot be kept because of the missing parallel path. This large value of V_l produces strong power oscillations during the failure of the line. They can be seen in Fig. 9.11. When the autonomous control system is in use, fuzzy module 1 for the generic rule 2 in UPFC 1 reacts immediately by deactivating the power flow controllers. The effect is visible in Fig. 9.10 with the output of the fuzzy module and the resulting outputs of the power flow controllers. The two power flow PI-controllers of UPFC 1 only cause a small increase of the manipulated variables during the failure. Consequently, the oscillations in the system occur more weakly during the failure than without the application of the autonomous control system.

It has to be mentioned that the shown effect only results from the slowing down of the controllers in order to prevent the false controlling effect. No FACTS damping controllers are present within that system. The damping could even be further improved if FACTS damping controllers were used.

Table 9.2. Result of the automatic topology analysis

FACTS device	UPFC 1	UPFC 2	UPFC 3
control path (node numbers)	18-5-6-7	7-9-10-11	14-15-16-17
parallel paths (node numbers)	18-17-7	7-19-14-12-11 7-19-11	14-19-7-17 14-12-11-19-7-17

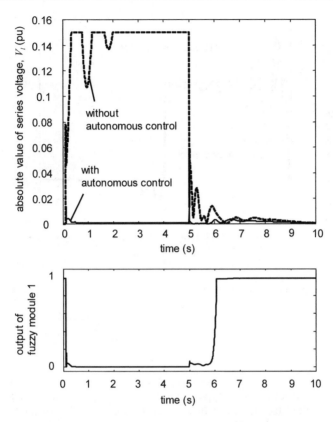

Fig. 9.10. Series voltage of UPFC 1 (above) and output of fuzzy module 1 of UPFC 1

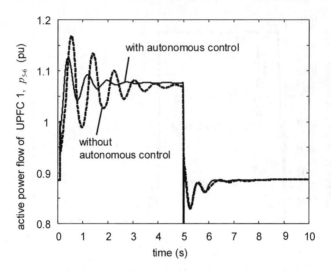

Fig. 9.11. Active power flow over UPFC 1

9.6.2 Increase of the Load

After an increase of the system loads the primary controllers of the power plants operate in order to cover the supplementary power requirement. Independent of this the three UPFC fix the power flows over the control paths constant with their fast power flow controllers. Consequently, they can for this moment not be used for the transmission of primary control power.

The capacity of line L_{11-19} is used to approximately 94 % before the load increase. A sloping increase of the system loads of 14 % happens at $t = 1s$. The primary control power generated by G4 has to be transmitted to the load at node 8 e.g. over line L_{11-19}, since the control of UPFC 2 first keeps the unchanged setpoint values. Without the autonomous control system a non-permissible overloading of L_{11-19} occurs (see Fig. 9.12). If this condition continued, a tripping of the transmission line would be inevitable. In the topology analysis, which has been executed before by the activated autonomous control system, this line is recognized as an element of a parallel path to UPFC 2, so that its integral-action controller counteracts directly on the line overloading and increases the setpoint value for the active power flow (see Fig. 9.12, below). This causes a shift of the power flow and a relief of L_{11-19}, so that its maximum loading limit of 0.62 pu (active power) can be kept (see Fig. 9.12, above). At $t = 1000$ s the loads are decreased to their original values, which means that the reason for the overloading has now disappeared. This has been simulated to show that the autonomous control system is able to reset itself when its coordinating control actions are not needed any more. The integral-action controller reduces its output back to zero.

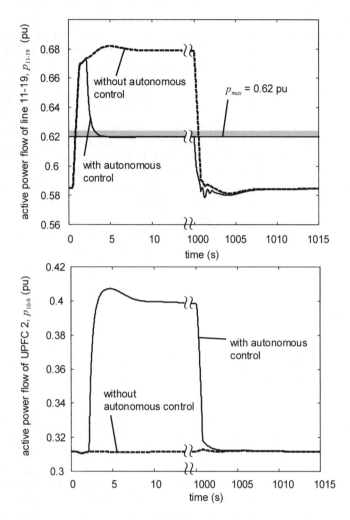

Fig. 9.12. Active power flow over line 11-19 (above) and active power flow over UPFC 2 (below)

9.7 Conclusions

The use of FACTS devices offers a flexible management of network-operation from a technical and economical point of view. Beyond this they cause many negative effects, which occur due to their short response time after critical events. Hence their advantages can only be used if automatic, quick, intelligent, and preventive coordinating measures are performed to eliminate those negative effects. The theory of autonomous systems offers a structural approach for a coordinating control system for FACTS devices. The necessary coordinating measures, which have to be realized by an autonomous control system, can be formulated as four generic rules. An autonomous control system has been developed and implemented for such a coordination so that the steady-state and dynamical system security is guaranteed after critical events. It automatically realizes the four generic rules for every operation condition of a power system. The control system specifies the generic rules within several steps and on different control levels so that concrete information is made available for decentralized autonomous components. For this, several techniques of Computational Intelligence, such as Fuzzy Control and Simulated Annealing, as well as conventional control techniques are applied. The concrete information consists e.g. of fuzzy rule bases and damping controller parameters and is generated preventively. Hence, the reaction of the system is as fast as possible and correct for the present operating condition of the power system. Simulation results have shown the improvements of the dynamical behavior and the secure steady-state operation, which are provided by the developed autonomous control system.

References

[1] Hingorani, NG (1993) Flexible AC Transmission. IEEE Spectrum, pp 40-45
[2] Edris A, Gyugyi L, et al (1999) AEP Unified Power Flow Controller Performance. IEEE Transactions on Power Delivery, vol 14, no 4, pp 1374-1381
[3] Povh, D (2000) Use of HVDC and FACTS. Proceedings of the IEEE, vol 88, no 2, pp 235-245
[4] Becker C (2000) Autonome Systeme zur koordinierenden Regelung von FACTS-Geräten - Autonomous Systems for Coordinating Control of FACTS Devices (in German). Doctoral Thesis, University of Dortmund, http://eldorado.uni-dortmund.de:8080/fb8/ls4/forschung/2001/becker
[5] Handschin E, Hoffmann W (1992) Integration of an Expert System for Security Assessment into an Energy Management System. Electrical Power & Energy Systems, vol 14
[6] Antsaklis PJ, Passino KM (eds) (1993) An Introduction to Intelligent and Autonomous Control. Kluwer Academic Publishers, Boston, Dordrecht, London
[7] Lei X, Lerch E, Povh D, Wang X, Haubrich HJ (1996) Global Settings of FACTS-Controllers in Power Systems. Cigré Session, Paper 14-305
[8] Larsen EV, Sanchez-Gasca JJ, Chow HH (1995) Concepts for Design of FACTS-Controllers to Damp Power Swings. IEEE Transactions on Power Systems, vol 10, no 2, pp 984-956
[9] Song YH, Johns AT (eds) (1999) Flexible AC Transmission Systems (FACTS), The Institution of Electrical Engineers, London
[10] Gyugyi, L (1992) Unified Power-Flow Control Concept for Flexible AC Transmission Systems. IEE Proceedings GTD, vol 139, no 4, pp 323-331

[11] Cigré Task Force 38.01.08 (1999) Modeling of Power Electronics Equipment (FACTS) in Load Flow and Stability Programs. Technical Brochure, Cigré SC 38, WG 01.08, Ref. No. 145

[12] Handschin E, Lehmköster C (1999) Optimal Power Flow for Deregulated Systems with FACTS Devices. Proc. of 13th PSCC, Trondheim, Norway

[13] Becker C, Rehtanz C (1998) Autonomous Systems for Intelligent Coordinated Control of FACTS Devices. Proc. of Bulk Power System Dynamics and Control IV - Restructuring, Santorini, Greece, pp 691-697

[14] Jungnickel D (1994) Graphen, Netzwerke und Algorithmen (in German), 3rd Edition. BI Wissenschaftsverlag, Mannheim, Leipzig, Wien, Zürich

[15] Wood AJ, Wollenberg BF (1996) Power Generation, Operation and Control, 2nd Edition. John Wiley & Sons Inc., New York

[16] Chen XR, Pahalawaththa NC, Annakkage UD (1997) Design of Multiple FACTS Damping controllers. Proc. of International Power Engineering Conference IPEC 1997, Singapur, pp 331-336

[17] King, RE (1999) Computational Intelligence in Control Engineering. Control Engineering Series, Marcel Dekker, Inc., New York, Basel

10 Multi-Agent Coordination for Secondary Voltage Control

Haifeng Wang

University of Bath, Bath, U.K.

This chapter introduces an application study of multi-agent coordination for the secondary voltage control in power system contingencies. The objective of the secondary voltage control is to eliminate any voltage violations in power system contingent operation. The introduction is presented through two example power systems, a simple single-machine infinite-bus power system and the New England 10-machine power system. The secondary voltage control is implemented through two types of FACTS (Flexible AC Transmission Systems) voltage controllers, SVC (Static Var Compensator) and STATCOM (Static Synchronous Compensator).

10.1 Introduction

As it is well known, power system voltage control has a hierarchy structure with two or three levels. At the primary level of the voltage control, the control devices attempt to compensate rapid and random voltage variations by maintaining their output variables close to the setting reference values. In some power systems, there can be a secondary level of the voltage control, which is designed to handle slow and large voltage variations, such as those produced by the hourly evolution of the load. The secondary voltage control utilizes the regional information. At the highest level, i.e., the tertiary voltage control, global information of the power system is used and the control is implemented by solving some optimization problems.

Previously, on-line power system voltage control was mainly about the primary voltage control implemented by the generator AVRs (Automatic Voltage Regulators). After several serious incidents of voltage collapse happened in the world, new schemes for the on-line voltage control and reactive power management in power systems have been pursued vigorously to improve power system voltage stability. One of the successful measures proposed is the secondary voltage control, initiated by the French electricity company, EDF, followed by some other electricity utilities in European countries [1]-[6]. The secondary voltage control closes the control loop of the reference value setting of the controllers at the primary level. So far, the research and applications of power system secondary voltage control has mainly been implemented by AVRs and the purpose has been focused on the improvement of power system voltage stability [1]-[6].

Although the hierarchical voltage control (primary/secondary/tertiary) may not be applicable to or fit well with the requirements of the reactive power and voltage management of *all* practical power transmission systems, the useful concept of the regional secondary voltage control could find more applications besides the improvement of system voltage stability. In this chapter, the implementation of the secondary voltage control by FACTS voltage controllers is introduced. The objective is to eliminate voltage violations in power system contingencies.

However, FACTS devices (SVCs and STATCOMs) are local controllers and secondary voltage control is a regional or area control operation, which needs the coordination of multiple SVCs and STATCOMs. This coordination is difficult to be implemented by decentralized actions of individual controllers without their mutual communication.

On the other hand, if a conventional centralized operation for the secondary voltage control is used, it involves all the controllers at all time. This is neither necessary nor efficient, because in the power system, only the controllers, which are electrically close to each other, can collaborate to provide support to each other. Hence, this chapter introduces the secondary voltage control by FACTS devices based on the multi-agent collaboration.

Multi-agent theory belongs to the Distributed Artificial Intelligence (DAI), which has been developed and mainly applied for constructing large, complex and knowledge-rich software systems [7]. DAI advocates the problem solving by a number of semi-autonomous agents, which communicate and collaborate with each other. From the point of view of system control, a multi-agent system is different to both the centralized and the decentralized control. The multi-agent system works in a decentralized control regime. However, it requires the communication and cooperation if found necessary, not among all the agents but only *between* closely related agents with common interests.

The fundamental collaboration paradigm of the multi-agent system lies in the task sharing. Therefore, multi-agent collaboration fits well the requirement of the secondary voltage control by FACTS voltage controllers (SVCs and STATCOMs). The collaboration is not among all but between these SVCs and STATCOMs, which contribute to the maintenance of voltage profile in their own multi-agent system and can provide reactive power support to each other.

10.2 Multi-Agent Voltage Management – Feasibility Study

10.2.1 Necessity of the Secondary Voltage Control in Power System Contingencies

This section demonstrates the feasibility study of multi-agent collaboration for the coordinated secondary voltage control in power system contingencies. An individual power system voltage controller, such as AVR, SVC or STATCOM, is treated as an agent. Those electrically closely related voltage controllers in a region of a power system form a multi-agent system, working together for the management of regional voltage profile. Task sharing in the multi-agent system is achieved by either communication or local estimation for a common objective of maintaining the regional voltage profile. In the power system, there could be many multi-agent systems formed and each voltage controller could be the agent in more than one multi-agent system. The conflict of the actions taken by individual agent for the task sharing of different multi-agent systems, which it joins is avoided, because each agent is semi-autonomous and acts on the basis of self-interested cooperation. The communication between different multi-agent systems is unnecessary, because the power system voltage controllers electrically far away, which are the agents belonging to different multi-agent systems, have little influence on each other as far as the voltage control through reactive power regulation of the power system is concerned.

The feasibility study is demonstrated as follows by a simple example power system installed with an SVC and a STATCOM as shown by Fig.10.1.

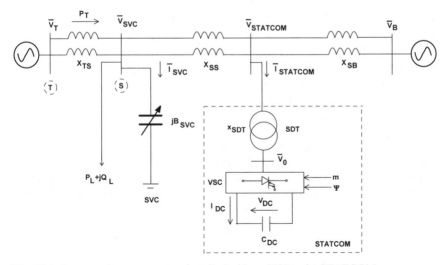

Fig. 10.1. An example power system installed with an SVC and a STATCOM

This figure is a system of a power plant sending power through two parallel long transmission lines to a large system and a regional load center connected at the SVC substation. The parameters of the power system are (in per unit):

$$H = 4.0s., D = 4.0, T_{d0}' = 5.044s., x_d = 1.0, x_q = 0.6,$$
$$x_d' = 0.3, x_{st} = 0.4, x_{ss} = 0.3, x_{ss} = 0.3$$
$$x_{SDT} = 0.15, x_{svcl} = 1.0, x_{svcc} = 1.0$$
$$K_A = 10.0, T_A = 0.01s., T_C = 0.05s., C_{DC} = 1.0, V_{DC0} = 1.0$$
(10.1)

The STATCOM consists of a step-down transformer (SDT) with a leakage reactance x_{SDT}, a three-phase GTO based voltage source converter (VSC) and a DC capacitor. The enhanced dynamic mode of the STATCOM is [8]:

$$\overline{V}_0 = mV_{DC}(\cos\psi + j\sin\psi) = mV_{DC}\angle\psi$$
$$\frac{dV_{DC}}{dt} = \frac{m}{C_{DC}}(I_{L0d}\cos\psi + I_{L0q}\sin\psi)$$
$$\overline{I}_{L0} = I_{L0d} + jI_{L0q}$$
(10.2)

where m and ψ are the modulation ratio and phase of the VSC respectively. The VSC generates a controllable AC voltage source $v_0(t) = V_0\sin(\omega t - \psi)$ behind the leakage reactance. The difference of the voltage magnitude between the STATCOM bus AC voltage, $\overline{V}_{STATCOM}$, and $\overline{V}_0$ produces the reactive power exchange between the STATCOM and the power system, which is controlled by modulating m, so as to regulate the AC voltage at the STATCOM busbar. The active power injection from the power system into the STATCOM is controlled to keep a constant DC voltage across the DC capacitor by modulating ψ. Fig.10.2 shows the dynamic model of the STATCOM AC and DC voltage regulators [8].

The SVC installed in the example power system of Fig.10.1 is a Thyristor-Controlled-Reactor-Fixed-Capacitor (TCR-FC) type of SVC. For the TCR-FC type of SVC, the equivalent admittance is [9]:

$$jB_{SVC} = [1 - C(\alpha)] / jX_{SVCL} - 1 / jX_{SVCC}$$
$$C(\alpha) = (2\alpha - \sin 2\alpha) / \pi - 1$$
(10.3)

where α is the firing angle of the firing circuit of the thyristors. Fig.10.3 shows the configuration of the SVC voltage controller.

The following contains two scenarios of system contingencies to demonstrate the necessity of the secondary voltage control of AVR on the generator, the SVC and STATCOM installed in the power system.

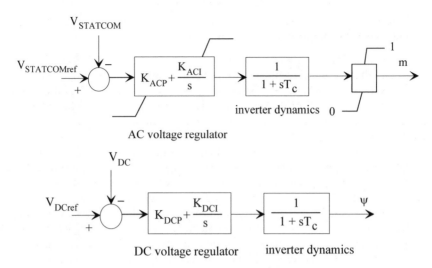

Fig. 10.2. Configuration of the STATCOM voltage controllers

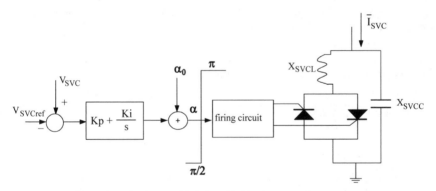

Fig. 10.3. Configuration of TCR-FC SVC installed in the example power system

- Scenario 1:   The load flow condition of the power system was:
  $P_T = 1.14\,\text{pu}$, $V_T = 1.05\,\text{pu}$, $V_{SVC} = 1.0\,\text{pu}$, $V_{STATCOM} = 1.0\,\text{pu}$, $V_B = 1.0\,\text{pu}$
  At 1.0 second, one of two transmission lines connecting bus T and S was lost
  together with the loss of half of the load connected at the SVC substation. At
  15.0 seconds, the supply to the lost half load was recovered without the reconnection of the lost transmission line.
- Scenario 2:   The load flow condition of the power system was:
  $P_T = 1.21\,\text{pu}$, , $V_T = 1.0\,\text{pu}$, $V_{SVC} = 1.0\,\text{pu}$, $V_{STATCOM} = 1.02\,\text{pu}$, $V_B = 1.0\,\text{pu}$.
  At 1.0 second of the simulation, a three-phase short circuit occurred on the

transmission lines connecting bus T and S. The fault was cleared after 120ms with one of two transmission lines out of service.

Fig.10.4 shows the simulation results of the bus voltage at the SVC substation. From Fig.10.4 it can be seen that during the abnormal contingent operation of the power system, the voltage level at bus S where the SVC is installed dropped below the allowable limit of 0.95 pu. The reason is that the SVC reached its limit of reactive power supply and was not able to provide more support to the voltage at bus S as required.

In order to eliminate the voltage violation at bus S in the contingent operation of the power system, extra reactive power supply is needed from the adjacent reactive power resources, which can be obtained from the actions of the generator AVR or/and the STATCOM in the example power system. This action can be activated through the secondary voltage control of the AVR and the STATCOM.

The results of two scenarios demonstrate the necessity of the secondary voltage control of power system voltage controllers involved for the voltage management during system contingencies. Although in this fairly simple example power system, the voltage violations in two scenarios demonstrated above could be avoided by careful operation planning, in more complicated practical systems, voltage management during system contingencies, such as the voltage recovering from a major outage, voltage management following a partial system collapse, unseen abnormal operation conditions or even a sudden failure of a reactive power resource, cannot be covered fully by system planning. Through on-line secondary voltage control of closely related voltage controllers installed in power systems, the voltage management during system contingencies is possible.

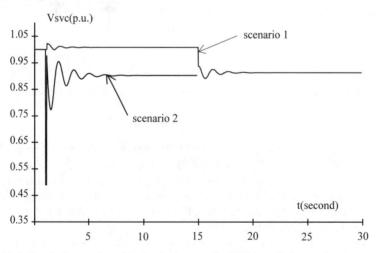

Fig. 10.4. Simulation results of the bus voltage at the SVC installation station

10.2.2 Multi-Agent Collaboration to Eliminate Voltage Violations

There are three voltage controllers in the example power system of Fig.10.1, the generator AVR, the SVC and the STATCOM. These three agents could form one agent system. However, considering that the influence between the generator AVR and the STATCOM is small, here two agent systems are formed. They are:

- AS1, joined by the generator AVR and the SVC;
- AS2, joined by the SVC and STATCOM.

The conflict of the action requirements between these two agent systems is avoided by defining the following self-interests in an order of priority:

- Physical limitation on the action of the agents, such as the maximum output of reactive power,
- No voltage violation on the bus where the agent locates.

More self-interests could be added for each different agent if found necessary. The action of each agent will never jeopardize its self-interests. The operation of all agents in an agent system is for a common objective to minimize the voltage violation in the power system.

For the example power system with two agent systems, the following two schemes can be applied to implement the coordinated secondary voltage control.

- Scheme 1: Multi-agent coordination of task sharing based on the communication. The collaboration among agents can be based on the type of "request and response" protocol as shown by Fig.10.5. Each agent monitors the voltage level at its location and once it finds the voltage violation, it will activate its reactive power reserve to sort out the violation by itself first. If this fails, it will select agents in its agent system to send out the request to perform the task to eliminate the voltage violation as an originator of the task. The agents receiving the request will accept the request if it is found that there is no conflict to their self-interests. They will perform the task, which is the action of the secondary voltage control by changing the reference setting of their primary voltage control. Then they will send response to the originator to inform of the action taken. This "request and response" protocol could be completed in one or more cycles until the voltage violation is eliminated. If no task response is received by the originator and the voltage violation still exists, it means the voltage violation cannot be solved by the secondary voltage control.

- Scheme 2: Multi-agent coordination of task sharing based on the local estimation. The collaboration among agents is the type of "estimation and voluntary action" as shown by Fig.10.6. The implementation of this scheme relies on the configuration of the power system. The electrical connections among agents in the example power system make it possible to estimate the voltage level at other agents' location from the measurement at each agent's location. For example, through the measurement of the bus voltage and line current at the STATCOM bus, the voltage at the SVC bus can be calculated.

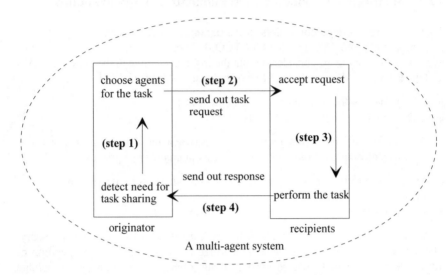

Fig. 10.5. Multi-agent collaboration based on the "request and response" protocol

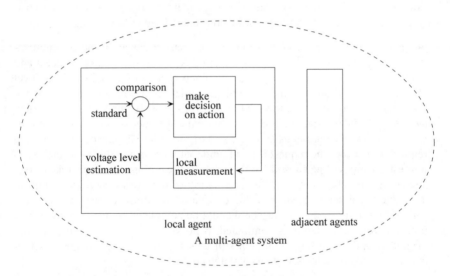

Fig. 10.6. Multi-agent coordination based on the local estimation

Therefore, the application of this scheme does not need the communication between agents. Each agent monitors the voltage level at not only its location but also the locations of adjacent agents from the measurement of its bus voltage and line current. Once the voltage violation is detected, no matter where it may be, the agent makes decision to change the reactive power supply through its secondary voltage control to eliminate the voltage violation. This is a voluntary action without jeopardizing its self-interests.

Those two scheme of multi-agent coordination were applied for the example power system and tested for the secondary voltage control in two scenarios presented in the previous section. The sequence of events and agent actions of applying scheme 1 and 2 respectively for scenario 1 is

1. The load flow condition of the power system was: $P_T = 1.14\,\text{pu}$, $V_T = 1.05\,\text{pu}$, $V_{SVC} = 1.0\,\text{pu}$, $V_{STATCOM} = 1.0\,\text{pu}$, $V_B = 1.0\,\text{pu}$. At 1.0 second, one of two transmission lines connecting bus T and S was lost together with the loss of half of the load at the SVC substation.
2. At 15.0 seconds, the supply to the lost half load was recovered without the reconnection of the lost transmission line.
3. When scheme 1 was implemented, at 16.0 seconds, the voltage violation was detected by agent SVC and it cannot be eliminated by itself. Then agent SVC decided to send task sharing request to agent AVR and STATCOM. After receiving the request, agent AVR decided not to respond because it was operating at the maximum voltage level allowed (1.05 pu) and to increase its reactive power output would damage its self-interests. While agent STATCOM accepted the request, performed the task by increasing its voltage reference setting to 1.05 pu at 20.0 second and sent the response to agent SVC.
4. When scheme 2 was implemented, the voltage violation at SVC bus bar was detected by both agent AVR and STATCOM at 16.0 seconds. Agent AVR did not act considering its self-interests. Agent STATCOM held for 4 seconds to wait for the action from agent SVC itself and finally decided to provide voluntary support by increasing its voltage reference setting to 1.05 pu at 20.0 seconds.

Fig.10.7 presents the simulation results of the multi-agent coordination of secondary voltage control for scenario 1. For scenario 2, the sequence of events and agent actions of applying scheme 1 and 2 respectively is

1. The load flow condition of the power system was: $P_T = 1.21\,\text{pu}$, $V_T = 1.0\,\text{pu}$, $V_{SVC} = 1.0\,\text{pu}$, $V_{STATCOM} = 1.02\,\text{pu}$, $V_B = 1.0\,\text{pu}$. At 1.0 second of the simulation, a three-phase short circuit occurred on the transmission lines connecting bus T and S. The fault was cleared after 120 ms with one of two transmission lines out of service.
2. When scheme 1 was implemented, at 5.0 seconds, the voltage violation was found by agent SVC and it cannot be removed by itself. Then agent SVC decided to send task sharing request to agent AVR and STATCOM. After receiving the request, agent AVR decided to respond by increasing its voltage reference setting to 1.05 pu at 8.0 seconds and sent the response to agent SVC. While agent STATCOM held the response because it considered that its current voltage level at 1.02 pu might have helped agent SVC. At 9.0 seconds a second request was sent by agent SVC to agent STATCOM and agent STATCOM decided a quick response at 10.0 seconds by increasing its voltage reference setting to 1.05 pu.

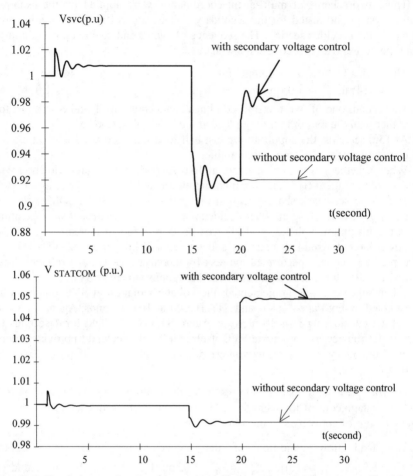

Fig. 10.7. Simulation results of multi-agent coordination of secondary voltage control for scenario 1

3. When scheme 2 was implemented, the voltage violation at SVC bus bar was detected by both agent AVR and STATCOM at 5.0 seconds. Agent AVR held the action for 3 seconds waiting for the self regulation by agent SVC and decided to provide support to agent SVC by increasing its voltage reference setting to 1.05 pu at 8.0 seconds. While agent STATCOM held the action and found the increase of SVC bus voltage. However, it found again that the voltage violation was not removed and finally decided to provide voluntary support to increase its voltage reference setting to 1.05 pu at 10.0 seconds.

Fig.10.8 presents the simulation results of the multi-agent coordination of secondary voltage control for scenario 2.

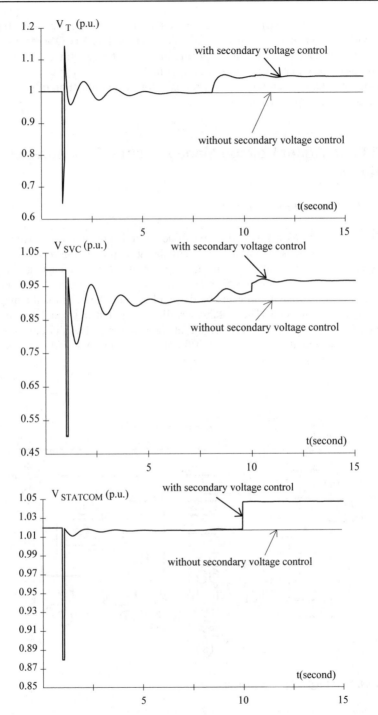

Fig. 10.8. Simulation results of multi-agent coordination of secondary voltage control for scenario 2

Fig.10.7 and 10.8 confirm the possibility that the multi-agent coordination for the secondary voltage control involving AVR, SVC and STATCOM is very effective for the management of voltage profile during system contingencies to eliminate the voltage violation in the power system. In the following section, a better multi-agent collaboration protocol is proposed for the coordination of more agents in a large complex power system.

10.3 Multi-Agent Voltage Management – Collaboration Protocol

10.3.1 Collaboration Protocol

Fig.10.9 shows the New England 10-machine 39-node power system. This system will be used to demonstrat a multi-agent collaboration protocol in this section. The voltage profile of the power system can be given as a three-dimensional graph of Fig.10.10, where the horizontal plane of the graph represents the electric connections of the system indicated by dashed lines and the height is the magnitue of voltage at each node. The voltage profile along transmission lines are also demonstarted in Fig.10.10. The voltage at the horizontal level is 0.95 pu. Two SVCs are installed at node 4 and 13 (B04 and B13) and two STATCOMs are at node 8 and 14 (B08 and B14) respectively.

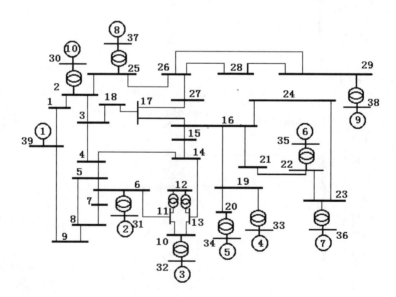

Fig. 10.9. New England power system

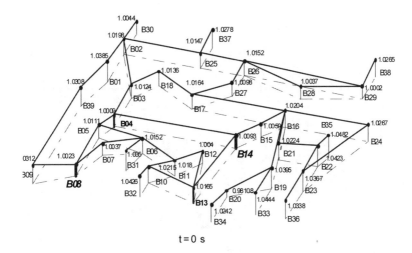

$t = 0$ s

Fig. 10.10. Voltage profile of the New England system

They form a multi-agent system with the first SVC at node 4 being agent 1(AG1), the first STATCOM at node 8 AG2, the second SVC at node 13 AG3 and the second STATCOM at node 14 AG4. The coordination among the agents in the multi-agent system is demonstrated in the previous section to eliminate the voltage violations at these locations where agents reside via the secondary voltage control. However, in a complex power system, this idea of the secondary voltage support only to the nodes where the agents locate is not appropriate any more, which can be illustrated by the following two scenarios of system contingencies.

- Scenario 1: After a 20% increase of load at node 15, the transmission line between node 15 and 16 was lost. The voltage profile of the power system at this operation of system contingency is shown by Fig.10.11, where the voltage violation appeared at node 15 (below 0.95 pu at B15) can be observed.
- Scenario 2: After a 25% increase of load at node 8, the transmission lines between node 5, 6 and 7 were lost. The voltage profile of the power system at this contingent operation mode is shown by Fig.10.12, which indicates the voltage violations at node 7 and 8 (below 0.95 pu at B07 and B08).

From Fig.10.11 and 10.12 it can be seen that in system contingent operations, voltage violations can appear not only at the nodes which are key nodes installed with voltage controllers (agents), but also at the adjacent nodes to the agents' locations. It is also possible that the voltage violations occur at the adjacent nodes to the agents when the voltage level at the agents' locations is normal. Since the agents' stations and associated intelligence are already in place, an extra function can be added for an agent to monitor the voltage level at the adjacent nodes, which are "reachable" from the agent via estimation by its local measurement (hence feasible).

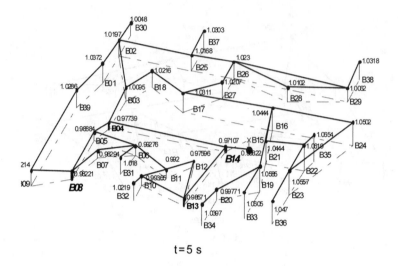

t = 5 s

Fig. 10.11. Voltage profile of the system in scenario 1

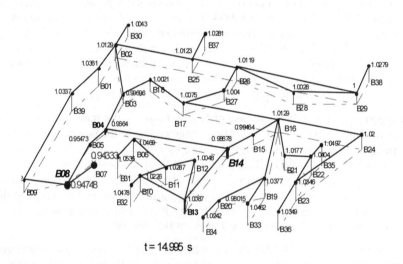

t = 14.995 s

Fig. 10.12. Voltage profile of the system in scenario 2

This extends the coverage of each agent in the system from a single node to its multiple nodes. Table 10.1 shows the arrangement of multi-node coverage of the secondary voltage control for each agent in the New England power system of Fig.10.9. Obviously this multi-node coverage extends the influence of the agents in the power system, which will be able to enhance the capability of the multi-agent system to eliminate the voltage violations and to achieve a better voltage management in system contingencies.

The extension of agents' coverage from single node to multiple nodes as shown by Table 1 results in the increase of agents' responsibility in system contingencies. However, in the same time, it increases the complexity of agents' coordination. To accommodate this extension, the simple request-reply type of coordination of multiple agents in system contingencies introduced in the previous section is improved to be a collaboration protocol as shown in graphical form of the infinite state machine by Fig.10.13.

Table 10.1. The arrangement of multi-point coverage for agents in Fig.10.9.

Device	SVC 1	STATCOM 1	SVC 2	STATCOM 2
Agent	AG1	AG2	AG3	AG4
Agent location	Node 4	Node 8	Node 13	Node 14
Adjacent nodes covered by the agents	Node 3 and 5	Node 5, 7 and 9	Node 10 and 12	Node 15
MULTICAST group	AG1, AG2, AG4	AG1, AG2	AG3, AG4	AG1, AG3, AG4

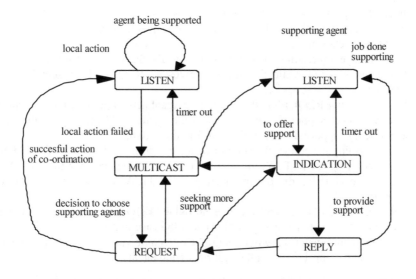

The multi -agent system

Fig. 10.13. Graphical finite state machine of the collaboration protocol of the multi-agent system

The finite state machine is a key concept used to specify and verify communication protocols [10]. With this technique, each protocol agent is always at a specific state at every instant of time. The state of the complete system is the combination of all states of protocol agents and channels, which are carriers of the transitions of the states. By doing reachability analysis [11], it can be determined if a protocol is correct or not. For simplicity, in Fig.10.13, only the agent being supported and one supporting agent are presented. From Fig.10.13 it can be seen that the collaboration protocol can be illustrated further as follows:

1. Each agent constantly monitors the voltage profile of its own node and adjacent nodes and the initial state of the agent is LISTEN;
2. Once an agent detects a voltage violation, it checks its capacity limit and acts on its own via the secondary voltage control if the limit permits. If the voltage violation disappears, the agent returns to the initial state. This is the loop of local action in Fig.10.13. Otherwise it sets up a timer and enters the status of MULTICAST, sending signal for help to its neighbor. The groups of MULTICAST are predetermined by examining the electric connections and locations of agents. Only agents who can support each other are put into a same group;
3. When an agent in LISTEN receives a message of MULTICAST for voltage support, it checks its capacity and the voltage profile of its own and adjacent nodes. This gives it an estimation of the amount of support it can offer. Then it sends back an INDICATION to the source of MULTICAST to tell its offer of support;
4. After the agent in MULTICAST receives several offers of INDICATION, it makes decision to choose the support from one agent. Then it sends REQUEST to the agent chosen for support;
5. After the agent intending to support sends out INDICATION, it sets up a timer. If the timer is out while the agent receives no REQUEST, it goes back to LISTEN;
6. Upon receiving REQUEST, the supporting agent acts to provide the amount of reactive power support it can and enters the status of REPLY by sending the agent being supported a REPLY to confirm the action;
7. After finishing the action, the supporting agent returns to the status of LISTEN;
8. After receiving the REPLY, the agent being supported goes back to LISTEN if the voltage violation disappears or enters MULTICAST if not to seek further support;
9. If the agent in MULTICAST receives no INDICATION after the timer is out, it goes back to the status of LISTEN. This is the case that the voltage violations in system contingencies are not able to be overcome via the coordinated secondary voltage control by the collaboration among agents.

In the collaboration protocol, two timers are assigned to prevent the deadlock where there is no progress can be made from a state.

10.3.2. Test Results by Simulation

For the New England power system of Fig.10.9, four groups of MULTICAST are formed initiated by four different agents shown in the last row of Table 10.1. Each SVC at node 4 and 13 was assigned a cell-immune-response secondary voltage controller [12] and each STATCOM at node 8 and 14 was equipped with a humoral-immune-response voltage controller [13]. These voltage controllers were designed by use of the approaches introduced in [12][13]. Extensive simulation has been carried out to verify the operation of multi-agent collaboration. The objective of simulation test is to demonstrate the effective elimination of voltage violations in system contingencies by the coordinated secondary voltage control. In the following, only four representative test results are presented. They are chosen to demonstrate different types of operation of the multi-agent system.

Test 1: At the 1st second of the simulation, the load at node 15 increased by 12%, which caused voltage drop at node 15, V_{15}. At the 2nd second, the transmission line between node 15 and 16 was lost which led to further drop of V_{15} below the constraint of 0.95 pu. This voltage violation at node 15 was noticed by AG4 and it evoked the action of the secondary voltage control of AG4 at the 3rd second. At the 10th second, the faulted transmission line was switched back successfully. The simulation result is shown by Fig.10.14. From Fig.10.14 we can see that during the contingent operation of the system from the 2nd second to 10th second, voltage violation at node 15 existed until at the 3rd second when it was eliminated effectively by the secondary voltage control of AG4. Fig.10.15 gives the graphical illustration of the finite-state machine of the operation of AG4. We can see that the operation was completed by local action of AG4 without involving any support from other agents in the system.

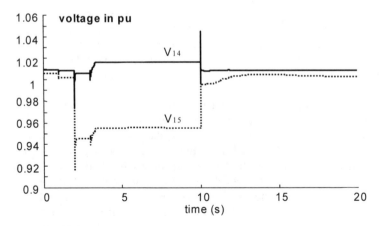

Fig. 10.14. Simulation result of test 1

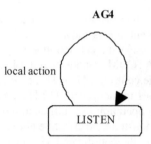

Fig. 10.15. Finite-state flow chart of test 1

Test 2: At the 1st second of the simulation, the load at node 15 increased by 20% and at the 5th second, the transmission line between node 15 and 16 was lost which led to the drop of V_{15} below the constraint of 0.95 pu This voltage violation at node 15 triggered the action of the secondary voltage control of AG4 immediately. However, the local action by AG4 failed to eliminate the voltage violation at node 15 because it has reached its upper limit of capacity. Hence at 6th second, it sent *multicast* for support to AG1 and AG3 in its *multicast* group. After the *indication* from AG1 and AG3 was received, AG4 made the decision to receive the support from AG3 and sent the *request* to AG3. At the 8th second, AG3 sent the *reply* to AG4 and took the action of the secondary voltage control. This action effectively eliminated the voltage violation at node 15. The simulation result is shown by Fig.10.16. From Fig.10.16 we can see that in the contingent operation of the system from the 5th second on, there was a voltage violation at node 15 from the 5th second to the 8th second.

After AG3 took the action to support AG4 at the 8th second, the voltage violation was eliminated effectively. The whole process of the agents' collaboration is shown by the finite state machine graph of Fig.10.17, where the sequence of agents' action is indicated by numbers. This is the demonstration of multi-agent operation with one agent being supported by another one after the failure of agent local action.

Test 3: At 0.2 second of the simulation, the load at node 15 increased by 20% and at the 2nd second of the simulation, the transmission line between node 15 and 16 was lost which led to the drop of V_{15} below the constraint of 0.95 pu. This voltage violation at node 15 evoked the action of the secondary voltage control by AG4 immediately. However, the location action by AG4 failed to eliminate the voltage violation at node 15, which became worse at the 5th second with the transmission line between nodes 10 and 13 was lost. At the 3rd second, AG4 sent *multicast* for support to AG1 and AG3 in its *multicast* group. This time AG4 decided to accept the support from AG1. At the 8th second, the support from AG1 took effect but failed to eliminate the voltage violation.

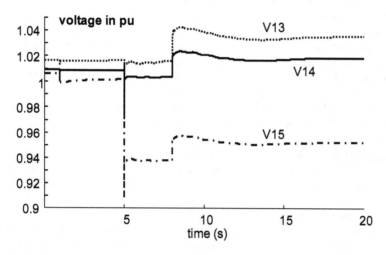

Fig. 10.16. Simulation result of test 2

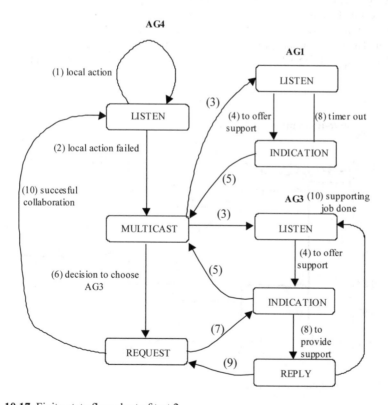

Fig. 10.17. Finite-state flow chart of test 2

Then *multicast* was sent to AG3 again, which brought AG3 into action. At 10th second, the action by AG3 successfully eliminated the voltage violation at node 15. The simulation result is shown by Fig.10.18. From Fig.10.18 we can see that during the contingent operation of the system from the 2nd second on, voltage violation existed until at the 10th second when AG3 was activated. With the joint support from AG1 and AG3, this voltage violation was eliminated at the 10th second. The whole process of the agents' collaboration is shown by the finite state machine graph of Fig.10.19, where the sequence of agents' action is indicated by numbers. This is the operation of multi-agent system with two agents supporting a third agent for the elimination of voltage violation detected in system contingency.

Test 4: At the 1st second of the simulation, the load at node 8 increased by 25% and at the 2nd second, the transmission line between node 6 and 7 was lost. At the 5th second, the transmission line between 5 and 6 was also lost which resulted in voltage violations at both node 7 and 8. These voltage violations at node 7 and 8 were confirmed by AG2 at node 8 at the 12th second of the simulation. However, the subsequent local action of the secondary voltage control by AG2 failed and it asked for support to AG1. The secondary voltage control by AG1 to respond to the request by AG2 took place at the 20th second of the simulation, which effectively eliminated the voltage violation at node 7 and 8. Fig.10.20 shows the simulation result. This is the test case where the voltage violations were at multiple locations, which were eliminated effectively by the coordinated secondary voltage control. The whole process of the agents' collaboration is shown by the finite state machine graph of Fig.10.21, where the sequence of agents' action is indicated by numbers.

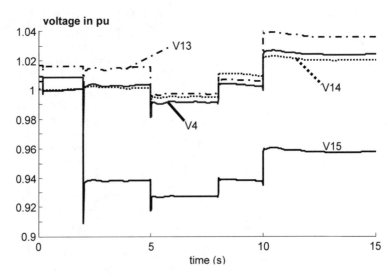

Fig. 10.18. Simulation result of test 3

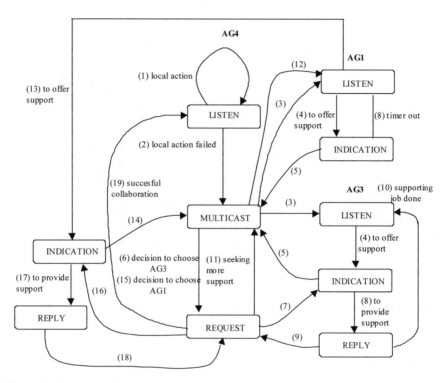

Fig. 10.19. Finite-state flow chart of test 3

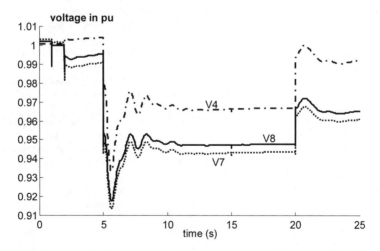

Fig. 10.20. Simulation result of test 4

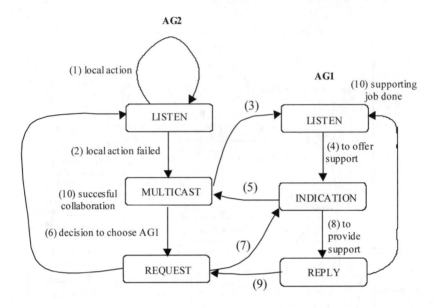

Fig. 10.21. Finite-state flow chart of test 4

References

[1]   Paul JP, Leost JY (1986) Improvements of the secondary voltage control in France. Proc. of IFAC Symposium on Power Systems and Power Plants Control, Beijing

[2]   Paul JP, Leost JY, Tesseron JM (1987) Survey of the secondary voltage control in France: present realization and investigation. IEEE Transactions on Power Systems, vol 2, no 1

[3]   Thorp JS, Ilic-Spong M (1984) Optimal secondary voltage control using pilot point information structure. Proc. of IEEE CDC, Las Vegas, USA

[4]   Lagonotte P, Sabonnadiere JC, Leost JY, Paul JP (1989) Structure analysis of the electrical system: application to secondary voltage control in France. IEEE Transactions on Power Systems, vol 4, no 2

[5]   Stancovic A, Ilic M, Maratukulam D (1991) Recent results in secondary voltage control of power systems. IEEE Transactions on Power Systems, vol 6, no1

[6]   Marinescu B, Bourles H (1999) Robust predictive control for the flexible coordinated secondary voltage control of large-scale power systems. IEEE Transactions on Power Systems, vol 14, no 4

[7]   Jennings NR (1994) Cooperation in industrial multi-agent systems. World Scientific.

[8]   Wang HF (1999) Phillips-Heffron model of power systems installed with STATCOM and applications. IEE Proc. Part C, no 5

[9]   Wang HF, Swift FJ (1996) The capability of the Static Var Compensator in damping power system oscillations. IEE Proc. Part C, no 4

[10]   Tanenbaum AS (1996) Computer Networking. Prentice-Hall International

[11] Lee CC (1990) Fuzzy Logic in Control Systems: Fuzzy Logic Controller, Part I-II. IEEE Transactions on Systems, Man and Cybernetics, vol 20, no 2

11 Agent Based Power System Visualization

Carsten Leder

Institute of Electric Power Systems, University of Dortmund, Germany

Many of the above described multi-agent solutions directly or indirectly support the control center staff within the difficult task of power system operation. Whereas intelligent agents solve subtasks autonomously, the supervision of the system state as a whole is still a human task. Therefore the user interface for energy management systems must be closely adapted to the human decision process.

This chapter presents an intelligent agent approach to power system visualization and control. The ability of intelligent agents to quickly adapt to new requirements makes them suitable for information provision in changing situations. Depending on the system state agents must decide which information is actually necessary and which visualization method is the most effective for its presentation.

11.1 Actual Problems in Power System Visualization

The result of the increasing electricity demand in fast developing regions of the world and of decreasing investments by grid companies in recently liberalized energy markets is an operation of the power system closer to technical limits. Furthermore frequently changing system states evoked by market transactions generate more and more situations, which the operators are not familiar with. These two main aspects make new solutions for the operators support in the control center essential. Fig. 11.1 shows the position of the new visualization concepts in relation to today's solutions. The well-established concepts are based on the device-focused visualization of measurements, messages and the grid topology [1]. The difficult task of interpreting the huge amount of data and extracting the essential information remains a human one. In order to support the operator in this task, recently published visualization concepts include the use of compact indicators for single aspects of the system state [2, 3] and information provision by innovative displays [4-6]. Examples for innovative displays are animated flow visualization [4] or matrix based representation of contingency analysis [5].

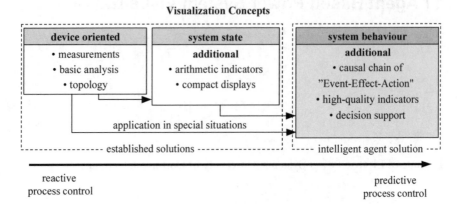

Fig. 11.1. Agent based visualization concepts extend established solutions

Today's visualization concepts display the system state and detailed aspects on the basis of single measurements. This is not sufficient to give an insight into system behavior, which is essential for a secure operation under the mentioned new boundary conditions. The control center staff is not well supported for the interpretation of separated pieces of information and the drawing of conclusion out of them for decision making.

To eliminate the deficits of existing solutions a new visualization concept will be introduced in the following. The purpose of the developed solution consists of the fact that a predictive process management instead of a reacting process control is implemented. This is not a matter of displaying the already known single measurements with new methods, but rather of visualizing the system behavior instead of the system state by showing process coherences, high-quality indicators and automatically generated control measures. The main focus lies on showing the causal chain of system behavior, which consists of 'Event-Effect-Action' and is described in detail in section 11.2.

While the existing visualization methods will continue being essential for routine jobs in the control center, the visualization concept presented here shall guarantee as a subsidiary system a predictive and secure operation, especially in critical situations. Within the new concept existing methods only come into operation when the presentation of topological and non-topological details is necessary for accurate analysis of partial aspects of the system behavior. Moreover these presented information goes beyond simply measurements (e.g. participation factors).

11.2 Decision Supporting Human-Machine Interface

The main idea of the new visualization concept is the representation of process coherences in a natural way, which actively supports the human decision process. The most complex functions of the human decision process 'interpretation' and 'planning' are partly transferred to an information management system. For a detailed physiological description of the human behavior in process control see [7]. Instead of understanding and visualizing the power system as a set of devices, the new concept is based on the interpretation as a causal 'event-effect-action'-chain [8].

11.2.1 Causality as the Natural Principle for Visualization

In general, every technical process can be described as an 'event-effect-action'-structure. Fig. 11.2 shows the causal chain of 'event', 'effect' and 'action', exemplified by the process of energy transmission. The event of a 'generator outage' leads to a certain effect. This effect can e.g. be the overloading of a transmission line. As a reaction to the modified and critical process conditions a control action must be taken, which in this case is a re-dispatch that reduces the overloading on the affected line.

A central task is to find out the causal dependencies between events, effects and actions. Strictly speaking, this for example means to answer the question which of two events is responsible for the voltage sag at a certain bus or which measure is sufficient to restore transient stability. The responsibility of an event for an effect is given by the simultaneous occurrence or must be determined by participation factors as e.g. generation shift distribution factors (GSDF). If an effect is detected, a specialized method, mostly based on Computational Intelligence (CI), is chosen which is individually adapted to the violated limit. This description points out that the diversity of tasks, which are strongly interweaved makes an intelligent agents solution obvious.

In order to detect events, quantify effects, and provide control actions powerful calculation tools are required. They must determine compact indicators as well as high-quality detail information.

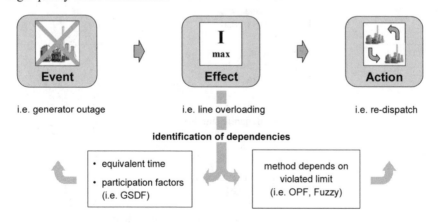

Fig. 11.2. Causal chain of technical process

Instead of data, information is shown. An overview of the events, effects and actions considered in the system is given in Table 11.1. The calculation of the required information with a multi-agent system is described in detail in section 11.2.3.

This approach is combined with the principle of 'means-end-hierarchy' which is described in [9]. In this case it is assumed that the process control is focused on reaching the optimum solution for a given task (e.g. the production of a good or the transmission of electrical energy). For this top aim further sub-objectives are formulated, whose degree of achievement are shown to the user. Based on these objectives links to functions and, finally, to the single means are created. The highly interactive character of the developed solution presented here meets these requirements.

11.2.2 Hierarchically Structured Information Provision

The ability of a human operator to analyze displayed information is limited to a maximum amount. Therefore the visualization concept must consider only the most essential information of a given system state. This is embedded in the concept in Fig. 11.3, which includes three hierarchically arranged visualization levels with different degree of abstraction.

Table 11.1. Considered aspects of power system behavior

Events	Outage of generator or transmission device
	Re-dispatch
	Reaching of generator limit (Q_{max})
	High load level / high load gradient
	Relevant forecast error
	Reaching of tapping limits
	Topology changes
	Large power transits
Effects	Thermal limit
	Voltage magnitude
	Frequency stability
	Fault level
	Voltage stability
	Static and dynamic stability
	(n-1) Security
Actions	Re-dispatch
	Generator voltage control
	Topology changes
	Transformer tapping
	Switching of reactive power compensation
	Load shedding / load control
	Adaptation of protection system
	FACTS

The constituents are the 'Global System View' (GSV), the 'Problem Specific Detail View' (PSV) and the 'Device Focused View' (DFV). The top level of the visualization concept is called GSV. By using higher-order system indicators a strong dimension reduction and - in connection with this - a better receptivity for the user is achieved. The display of system indicators according to the structure of the causal chain results in the comprehension of event, effect and necessary measures in their coherence. On this level the attention of the operator is directed to the critical objective of the process. The staff shall recognize a problem and thereupon request more information, which is assembled by the system dependent on the situation.

By selecting the respective indicator the user receives additional information in the second visualization level, the PSV. Based on the present system state the operator on this level will be enabled to analyze in detail the single system characteristics by the provision of the PSV. Thereby the structure 'event-effect-action' will be continued consequently. This means that continuative information will be made available for all three sub-areas, which goes beyond simple measurement values.

Thus, the display of the grid in varying abstraction levels is not sufficient, so that innovative presentation methods also come into operation for the visualization of the system behavior and the necessary measures (e.g. contour plot). In [10] the 'Congestion Clock' is presented as a human focused method for an efficient visualization of time- and place-dependent congestion data. Its orientation to well-known principles guarantees an accurate interpretation by the operator.

The special characteristic of PSV is the individual grouping of information according to the given operation problem. Through that it is considered that the required detail information is not independent of the violated limit.

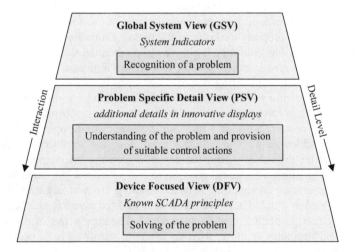

Fig. 11.3. Hierarchical structure of visualization concept

It is obvious that the analysis of e.g. voltage stability requires details about re-active power reserves and modal analysis, whereas for the examination of line overloadings current values and generation shift distribution factors are needed. Furthermore the network region from which information are given varies depending on the origin of the problem. Within this visualization level the operator is enabled to understand a problem and to solve it with appropriate methods. By choosing a suggested measure the access to the DFV takes place.

The DFV represents the lowest and thus the most detailed level of the visualization concept. It shows the network regions, which are relevant for control measures. The display of this level takes place after the manual activation of measures by the operator, but with the addition that the determination of the relevant region works automatically by means of the analysis of system indicators and heuristically determined spheres of influence. Thereby the direct topological environment of the affected devices is shown in the well-known single-line diagrams for sub-networks or substations. In this visualization level the operator solves the problem on the basis of the information obtained in the upper hierarchy levels.

Whereas the visualizations methods of DFV are well known from existing SCADA systems, GSV and PSV are described more detailed in the following sections.

11.2.3 Global System View and Problem Specific Detail View

The GSV presents the major aspects of the system state to the operator and allows him to realize the appearance of problems in advance. Higher-order indicators are calculated in order to solve this task [11-13]. The considered aspects are given in Table 11.1.

For the visualization of the indicators intuitive understandable pictograms are used. As the figure of a pictogram allows the operator to find a connection to the represented system property the color of the pictogram codes the magnitude of the assigned indicator. The colormap is adapted to established standards [14]. Hence a transition from a warning indicator value to a critical indicator value is represented by a color transition from yellow to red. In order to draw the attention of the operator to relevant information and to avoid the irritation by irrelevant information non-critical indicator values are colored by 'light grey' (inconspicuous) instead of 'green' (conspicuous). A single value can indicate 'normal operation', 'warning state' or 'forbidden state'. The indicator magnitude that separates one area from another must be individually adjusted depending on the belonging system property. In order to allow a sufficient reaction to critical events, the adjustment of the limits must guarantee enough time between entering the 'warning state' and entering the 'forbidden state'. Therefore the diversity of the dynamic characteristics of different system properties must be considered. The appearance of e.g. a voltage collapse needs minutes or more, whereas dynamic stability events occur in milliseconds. A screenshot of the GSV is shown in Fig. 11.4.

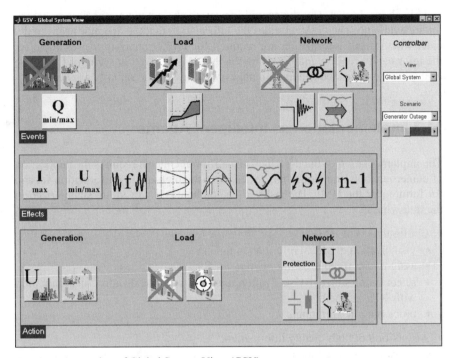

Fig. 11.4. Screenshot of Global System View (GSV)

The pictograms are vertically grouped by 'event', 'effect', 'action' and horizontally grouped by 'generation', 'load', 'network'. Whereas the principle of the causal process chain is applied for the vertical grouping, the horizontal grouping simplifies the navigation within the human-machine interface (HMI).

The symbols used for the pictograms must be discussed and optimized in usability tests with operators [15]. Their individual preferences, but also common standards, must be applied to find the optimal appearance, which minimizes misunderstandings and maximizes the efficiency of use. The meaning of the symbols in Fig. 11.4 is given for some example in the following list:

- The crossed out power plant (upper left corner) indicates a generator outage.
- The color of the nose curve (4th in 'effects'-frame) indicates voltage stability.
- The picture of the human operator and the switch (lower right corner) is for the suggestion of a topology change as control measure.

Each pictogram has got its own fixed position and is always visible, even in situations when the assigned indicator has not a critical value. This is necessary not to confuse the operator by frequently, substantially changing displays.

The application of a projection system in the control center allows the permanent display of the GSV and guarantees the availability for all operators. In a control center without a projection system the GSV is always visible on one monitor. In the case of a critical state change the staff's attention is drawn to the GSV by audible alarm. In that way the central indicators characterizing the system behavior are available. The GSV provides the information that the operator needs to decide, which detail information are required for further analysis. An example for the information request and provision via the HMI is given in section 11.4. In the example case, cascaded generator outages result in voltage stability problems.

11.3 Implementation as Intelligent Agents

The implementation of the described visualization concept requires the application of numerous tools and algorithms for creating the high-quality information set and for forming of the HMI. The tasks to be continuously solved are summarized in the following list:

- Configuration of HMI:
  - visualization of global information in GSV,
  - assignment of required information for PSV,
  - selection of suitable visualization methods and visualization of information with selected methods,
  - processing of user inputs.

- Acquisition of high-quality information:
  - analysis of measurements and message for the detection of events,
  - calculation of indicators to quantify effects on the system operation,
  - automatic determination of efficient control measures.

- Solving of global subtasks for information provision like e.g.:
  - load forecast,
  - state estimation.

- Solving of higher-order control tasks like e.g.:
  - reactive power voltage control,
  - load shedding.

- Solving of local control tasks like e.g.:
  - optimization of FACTS controller settings,
  - adaptation of protection parameters.

The complexity of the given tasks evidently shows that only the separation into autonomous components guarantees an efficient solving. It is also obvious, that the separated instances must adapt their behavior to the actual system state and must communicate with each other in order to determine the system state. Especially the last two aspects suggest the implementation as a multi-agent system. Fig. 11.5. presents examples for the required agents and their grouping in four classes.

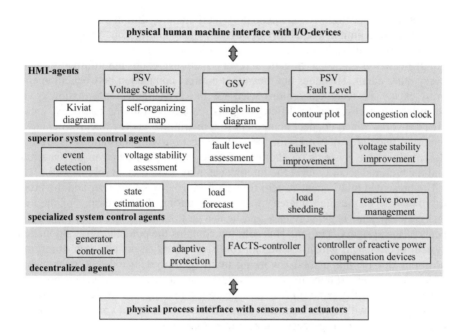

Fig. 11.5. Intelligent agents as bridge between operator and process

The required agents can be grouped into HMI-agents, superior system control agents, specialized system control agents and decentralized agents. The class of HMI-agents contains all components that configure the global and problem specific displays on the basis of the available information and request the generation of additional information. The intelligence lies basically in the ability to display relevant information and to ignore irrelevant.

The superior system control agents are able to extensively assess single operating limits through state indicators and further detail information. For the suggestion of corrective measures beside standard algorithms as well procedures of computational intelligence are used.

The superior system control agents are dependent on the results of specialized system control agents to reach their objectives. This group works closer to the process than the superior system control agents. The agents either take over the basic functions of data pre-processing (e.g. estimation) or transfer abstract instructions for the network management into concrete suggestions for switching operations or set-point commands. An example is the implementation of the command 'load shedding'. The component 'voltage stability improvement' generates the instruction 'load shedding of 100 MW in region west', whereupon the agent 'load shedding' must decide considering costs and contracts what loads is to be shed in the actual situation.

Decentralized agents carry out the local commands. The implemented control algorithms are adapted corresponding to the instructions of system control agents.

Furthermore decentralized agents report malfunctions during the execution of commands or a limited availability of devices, e.g. resulting from outages.

For reasons of readability the communication links between single agents are not presented in Fig. 11.5. In general, every agent can exchange information with every other agent. This does not mean that a direct connection between all agents must exist. The technical options used for the exchange of information cover software interfaces between different agents realized within one computer, local-area networks which connect computers in a distributed energy management system, and wide-area networks for the communication with substations as well as decentralized bus systems on station and field level. Dependent on the data rate and the required security level the Internet can play an important role as well.

The presented multi-agent system forms a bridge between the user and the process. Because of the high degree of autonomy during the processing of the sub tasks, the operator is given release and he is able to concentrate on the central aspects of the system operation. This solution contains the potential for a automated system operation in which the human role shifts from an acting to an observing position. But for this a robust working multi-agent system must be guaranteed. Moreover the opportunity for an intervention on the part of the human operator must remain. At the same time it must be considered that the process knowledge of the operator will be less trained because of the automatic network operation and that as a result human process management is complicated in case of a malfunction within the information management system.

To illustrate the theoretic explanations, section 11.4 describes how different agents cooperate within a multi-agent system to realize voltage stability assessment and improvement.

11.4 Verification

For the illustration of the described visualization concept it is applied to the network given in Fig. 11.6, which is a part of the UCTE system and consists of 126 nodes from which only a selection is shown in the figure. The chosen disturbance results from cascaded generator outages in node 2. Three uncorrelated reasons - a frozen sensor, an exploded circuit-breaker and a fire in the oil distribution system - cause the outage of three generators. As a result a lack of active and reactive power occurs in the western part of the system. Therefore a huge energy transport is required which leads to voltage stability problems and finally to a voltage collapse.

Discussions and analysis following the disturbance reveal one main reason for the breakdown. The information provided to the operator was not sufficient to analyze the problem adequately. The only available information were bus voltage, line flow and system messages. This did not allow the predictive analysis of the system behavior and the preventive activation of measures.

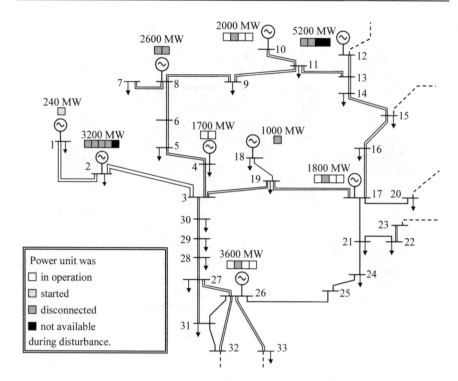

Fig. 11.6. Network for example case

11.4.1 User-Machine Interaction

For the example case a single instant is picked out to describe the interaction between the operator and the new visualization concept. Fig. 11.7 shows the appearance of the GSV for the moment just after the outage of the third generator. It can be seen that all three sections - event, effect, action - are affected. A colored version of the figures is available online in [16].

In the 'event'-section one symbol indicates a 'generator outage'. Furthermore the remaining reactive power reserve is reduced. Information about which generator broke down can be accessed by selecting the symbol and will be displayed in a separated window. In the same way the reactive power injection of relevant generators can be presented by bar graphs. Whereas the generator outage is less problematic in a light load situation, it is dangerous for peak load periods. Therefore the high load level is also indicated by an activated pictogram.

The described events result in critical effects for the system security. The voltage magnitude of at least one bus is reduced to the warning level. Much more dangerous is the decrement of voltage stability which can be seen from the assigned symbol showing the voltage stability indicator [11].

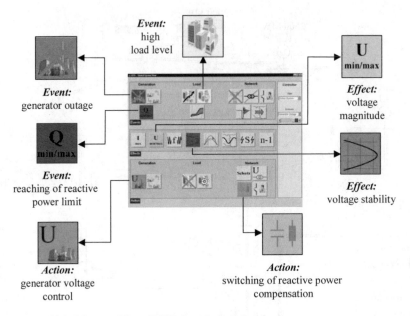

Fig. 11.7. Global System View (GSV) for selected situation

Because of the decreased voltage stability the visualization concept automatically presents the available control measures. In the given case the modification of the generator voltage setpoints and the switching of reactive power compensating devices are recommendable.

As mentioned in section 11.2 the function of the GSV is to make the operator recognize a problem within the power system. By indicating the weak voltage stability and the responsible events this task is solved. For ongoing analysis and the activation of control measures, further information are provided through the PSV (Fig. 11.8), which is available by selecting the 'voltage stability pictogram'.

The PSV for the analysis of voltage stability consists of three parts. The self-organizing map (SOM) (upper left corner) shows the distance of the actual operating point to the stability limit. The SOM is computational intelligence tool, which is used for online assessment and visualization of voltage stability [17, 18]. Each field represents, after a training process, a group of similar system states. The color indicates the voltage stability of the representative system state, which is calculated according to [11] during the training period. Beside the actual stability level, indicated by the black circle, numbers show the development of voltage stability in the latest periods. A very important detail information is the availability of reactive power from the generators. Therefore a bar graph in the lower left corner of Fig. 11.8 shows the injections and reserves of reactive power. The graph indicates that the remaining generators at node 2 have reached their reactive power limit.

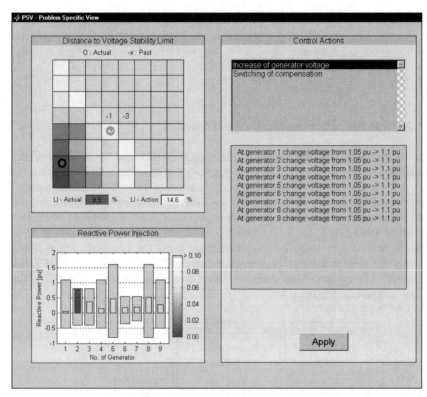

Fig. 11.8. Problem Specific Detail View (PSV) for selected situation

A major task of the PSV is the provision of control actions. Hence an expert system calculates a selection of control measures for the given situation. In the upper part of the 'control action'-frame a general description of the control measures is presented. A detailed description is available by selecting one of the alternatives and is displayed in the lower part of the frame. For the decision process the operator must know about the effect of a proposed control action on voltage stability. Therefore the green circle (light grey) at the SOM shows the prediction of the stability level, which is calculated by simulating the modified system state. Finally one of the proposed measures can be executed by selecting the 'apply'-button. This links the operator to the well-known sequences for switching and setpoint setting which he is familiar with.

From the description of the given example it is obvious that for the realization of the visualization concept requires the provision of high-quality information about the system behavior. Today's algorithms integrated in control center solutions are not sufficient, because of the following disadvantages:

- high computing time,
- difficult to maintain,
- low degree of transparency.

Therefore computational intelligence methods, namely neural networks and expert systems, are applied for the presented concept. In the last decade many approaches show good results for stability assessment and decision support using these solutions [19, 20]. As mentioned in section 11.3 the variety and the multitude of the required tasks suggest the realization within a multi-agent system [21].

11.4.2 Implementation as Multi-Agent System

To illustrate the multi-agent approach the implementation for the assessment, the correction and the visualization of voltage stability is described. To begin with, the design of a single intelligent agent for voltage stability assessment on the basis of the SOM will be described. Afterwards the different kinds of agents, which have to be combined for the predictive process management within one multi-agent system will be shown.

Fig. 11.9 shows the structure of an intelligent agent for the stability assessment including the three instances 'Organization Level', 'Coordination Level', 'Execution Level'. The agent's work starts off with the request of another agent (e.g. for decision support) or the operator.

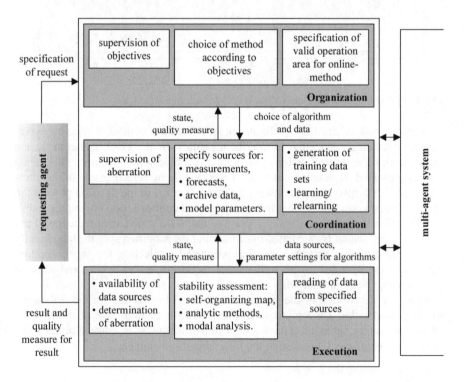

Fig. 11.9. Structure of agent for voltage stability assessment

The stability tool can be used not only for the assessment of the current system stability, but also for the assessment of control actions and for the planning purposes. Correspondingly, the request is specified concerning response time, result accuracy and level of detail.

One of the decisions in the organization level is the choice between the online algorithm for the calculation of the stability indicator and the more precise but slower analytic method. Whereas the SOM is the standard tool for operation because it is very fast, the second alternative is used for planning purposes. Furthermore the analytic tool must be used if the accuracy of the SOM is insufficient because of a discrepancy between the actual system state and the training states. In this case an adaptation of the training data set to the changed system states and the relearning of the SOM is required. If beside of the stability indicator additional detail information are needed, the modal analysis is activated. The organization level exchanges its decisions with the coordination level and controls its execution measured using a quality indicator.

The task of the coordination level is the specification of data requirements and data sources for the requested method. The analytic method requires actual measurements and system parameters. If the aberration of stability assessment is not sufficient, the generation of updated training data sets and the learning or relearning of the SOM are necessary. Furthermore the learning parameters are set and the success of the learning process is verified. A minimum mapping accuracy must be met. A successful learning is reported to the organization level. The readily trained SOM and the information about data sources are submitted to the execution level.

The chosen stability assessment method is applied in the execution level using the input data specified in the coordination level. A successful stability assessment is jeopardized if either the required data is not available or the aberration of the SOM is too big. The occurrence of such an error is reported to the superior level. The possible reactions to an insufficient result is the modification of the requirements or the negative reporting to the requesting agent.

After a successful solving of the subtasks, meeting the requested accuracy, the stability information and the reached aberration are submitted to the inquiring agent.

After exemplarily describing the internal structure of a single intelligent agent a multi-agent system is presented, which handles the task of voltage stability assessment and visualization. Fig. 11.10 shows the intelligent agents and the most essential communication relations. Communication in this context means either data submission or sending a request to another agent.

The central elements of the system are the superior system control agents. Their horizontal arrangement in Fig. 11.10 corresponds to the sequence of their activation during the analysis of a single situation. 'Event Detection' is responsible for the detection of events, which significantly influence the system state. If such an event occurs the assessment of voltage stability is required. The block 'Improvement of Voltage Stability' is activated if the stability margin is reduced to a dangerous level. For an adequate operation the superior system control agents need the assistance of various other agents.

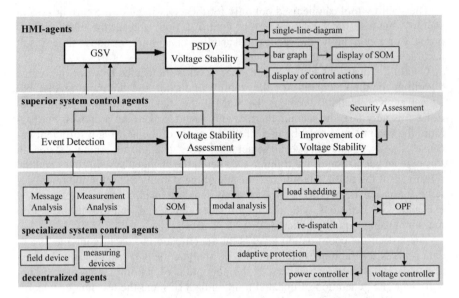

Fig. 11.10. Multi-agent system for voltage stability assessment and improvement

The event detection is based on messages and measurements from decentralized agents. Specialized system control agents filter the relevant data from the multitude of single values and recognize events, which influence voltage stability. Sources of messages and measurements are decentralized field devices or other specialized system control agents, which e.g. analyze the reactive power margins.

For the assessment of voltage stability the SOM is used as described above. The agent requires the grid topology, actual measurements and information generated by load shedding agents and re-dispatch agents. The last aspect is relevant because besides assigning stability indicators to the representative SOM states control measures can be assigned. Ongoing detail information is generated using the modal analysis.

The improvement of voltage stability is based on the adaptation of generally formulated control actions, e.g. blocking of tap changers, to the given situation. Therefore both results of the modal analysis and information of decentralized devices are required. Furthermore suggestions for load shedding and re-dispatch can be taken from the SOM. In order to prevent negative consequences for other technical limits the effect of stabilizing measures must be determined in advance. In this task a variety of agents is involved, which are displayed in a compressed way. The selected control actions are submitted to decentralized agents.

The information generated by the superior system control agents is sent to the HMI-agents in order to visualize it. Whereas indicators for events and the voltage stability are presented in the GSV, the detail information forms the basis for the PSV. Modal participation factors are shown in colored single line diagrams. Bar graphs display the available reactive power and the position of tap changers. The SOM and the control actions are visualized according to Fig. 11.8.

The described example illustrates how power system requirements are met with a multi-agent solution. In detail the following aspects must be mentioned:

- autonomous adaptation to new system states release the human operator and provides problem specific information,
- the variety of sub-problems is best solved by transparent autonomous agents,
- the self-monitoring character of the agents guarantees an adequate operation,
- the integration of CI-methods is the only way to solve problems like the automatic provision of control measures,
- the integration of CI-methods allows real-time operation.

11.5 Conclusions

The changing general conditions in power system operation are responsible for an urgent demand for new solutions for operator support. The described visualization concept supports both the operation of the power system closer to technical limits and the handling of frequently changing system states.

The multi-agent approach allows a quick adaptation to changing situations as well as the modular solving of subtasks in a transparent way. This guarantees a maximum assistance for the human operator. With the new visualization concept a change from a system focused approach to a human focused one is realized. The adaptation to the human decision process by showing the causal chain of process behavior increases the operator's knowledge about the system and reduces the danger of wrong reactions.

Beside using indicators for events, effects and actions in the global system view, the problem specific detail view gives ongoing information if required. Whereas today's HMI presents single data the new concept support the main functions of the human way of knowledge processing by providing high-quality information. This includes a predictive assessment of the power system security and stability as well as the provision of suitable control actions. The methods for information generation mainly base on Computational Intelligence. Many of them are document in recent publications.

The application of intelligent agents, which solve sub problems in power system operation autonomously still requires the involvement of human operators for centralized global tasks. The presented approach allows the optimal consideration of human skills in power system visualization.

References

[1]   Mahadev PM, Christie RD (1993) Envisioning power system data: concepts and a prototype system state representation. IEEE Trans. Power Systems, vol 8, pp 1084-1090

[2]   Hauser AJ, Verstege JF (1999) Seeing global power system states with compact visualisation techniques. In: Proc. of IEEE Power Tech '99, Budapest, Hungary, BPT 99-271-23

[3]   Sprenger W, Stelzner P, Schellstede G, Schäfer KF, Verstege J (1996) Compact and Operation Oriented Visualization of Complex System States. In: Proc. of CIGRE Session, Paris, France, pp 39-107

[4]   Overbye TJ, Weber RD (2001) Visualizing the Electric Grid. IEEE Spectrum , vol 38, pp 52-58

[5]   Bacher R (1995) Graphical Interaction and Visualization for the Analysis and Interpretation of Contingency Analysis Results. In: Proc. of 19th Power Industry Computer Application Conference (PICA), Salt Lake City, USA, pp 128-134

[6]   Avouris NM (2001) Abstractions for Operator Support in Energy Management Systems. Int. Journal of Electrical Power Energy Systems, vol 23, pp 333-341

[7]   Rasmussen J (1986) Information Processing and Human-Machine Interaction - An Approach to Cognitive Engineering. North-Holland, Elsevier Science Publishing, New York

[8]   Handschin E, Leder C (2001) Innovative Visualization of the Power System State for a Predictive Process Control. Electrical Engineering, vol 83, no 5/6, pp 297-301

[9]   Lind M (1991) Representations and Abstractions for Interface Design Using Multi-level Flow Modelling. In: Human-Computer Interaction and Complex Systems, Academic Press, London

[10]   Brosda J, Handschin E, Leder C (2002) Hierarchical Visualization of Network Congestions. In: Proc. of 14th Power Systems Computation Conference (PSCC), Sevilla, Spain

[11]   Lemaître C, Paul JP, Tesseron JM, Harmand Y, Zhao YS (1990) An Indicator of the Risk of Voltage Instability for Real-Time Control Applications. IEEE Trans. Power Systems, vol 5, pp 154-161

[12]   Handschin E, Leder C (2001) Automatic Decision Support with a new Visualization Concept for Power Systems. In: Proc. of IEEE Power Tech '01, Porto, Portugal

[13]   Kundur P. (1994) Power System Stabilitby and Control. McGraw-Hill, New York

[14]   IEC 60073 (2002) Basic and safety principles for man-machine interface, marking and identification - Coding principles for indication devices and actuators.

[15]   Nielsen J (1994) Usability Engineering. Morgan Kaufmann Publishers, San Francisco

[16]   Leder C. (2002) Visualisierungskonzepte für die Prozesslenkung elektrischer Energieübertragungssysteme. Dissertation University of Dortmund, http://eldorado.uni-dortmund.de:8080/FB8/ls4/forschung/2002/Leder

[17]   Leder C, Rehtanz C (2000) Stability Assessment of Electric Power Systems using Growing Neural Gas and self-organizing maps. In: Proc. of 8th European Symposium on Artificial Neural Networks, ESANN 2000, Bruges, Belgium

[18]   Kohonen T (1997) Self-organizing Maps. Springer-Verlag, Berlin, Germany

[19]   Rehtanz C (2002) Intelligente System in der elektrischen Energieversorgung. VDI Verlag, Düsseldorf

[20]   Wehenkel LA (1998) Automatic Learning Techniques in Power Systems. Kluwer Academic Publishers, Dordrecht

[21]   Amin M (2001) Toward Self-Healing Energy Infrastructure Systems. IEEE Computer Applications in Power, vol 14, no 1, pp. 20-28

12 New Applications of Multi-Agent System Technologies to Power Systems

Chen-Ching Liu, Hao Li, Yoshifumi Zoka

Department of Electrical Engineering, University of Washington, USA

This chapter provides a state-of-the-art summary of multi-agent system technologies for power system applications. New applications to controlled islanding of power systems and MicroGrid with distributed generations are discussed. A new defense system concept, Strategic Power Infrastructure Defense (SPID), has been developed. Controlled islanding is an important extension of the SPID system and the multi-agent system is a promising technology for implementation of the SPID concept. The University of Washington is also developing the intelligent control system for MicroGrids. In this chapter, the multi-agent technology applications to these two topics are discussed with specific examples.

The SPID program conducted by the Advanced Power Technologies (APT) Consortium was sponsored by Electric Power Research Institute and the U.S. Department of Defense (WO 8333-01). The MicroGrid research is sponsored by the U.S. National Science Foundation through grant, "Micro Grids with Distributed Generation in an Open Access Environment (ECS-0140629)."

12.1 Multi-Agent System Technologies

Over the past decades, many researchers and engineers have focused their efforts on intelligent system applications to various problem areas in power engineering. The reason for intelligent systems to attract their attention is that the technology has great potentials to handle large and complex systems. One of the important application areas for intelligent systems is the monitoring and control of complex systems such as power grids. A number of intelligent techniques have been applied to the analysis, control and decision making of power systems. Knowledge-based systems, neural networks, fuzzy logic, and other qualitative reasoning techniques have been developed [1]. However, in the competitive power industry, many non-utility entities are also participating in the open market. This makes the control scheme for power systems very different from the past and more complicated. This new industry structure also leads to a higher level of uncertainty in the planning and operation of the power grid. Most of conventional intelligent control systems are centralized and supervised in a top-down fashion. These systems might not be able to appropriately respond to rapid changes in the power systems. Note that the computational effort involves a large amount of information and data. In addition, maintenance of the existing intelligent control systems has become more important since the complexity and uncertainty of the power systems are increasing.

A promising technology to overcome the disadvantages is Distributed Artificial Intelligence (DAI). DAI has the abilities to meet the demand for automation and to tackle uncertainties of the systems. The tasks to control the power systems are divided into various modules in DAI systems; maintenance can be simplified relative to the conventional centralized structure since each module is designed to handle well-defined and limited tasks. Since the computational effort for each module is not high and each module can work independently, the overall computation can be carried out in parallel within a short time, allowing the system to perform in real-time. While each module has the criteria and knowledge to achieve its own goals, some major tasks, results, and decisions are shared by other modules to achieve the global goals. This DAI framework can achieve effective controls for unpredictable, uncertain systems through the integration of different views and problem-solving techniques in various modules.

Recent research in DAI has focused on the multi-agent system (MAS), which includes a number of agents coupled with networks and protocol for interaction among the agents. MAS can be regarded as a new design method to construct a complex system with "de-centralized" fashion in contrast to conventional centralized architecture. Each agent can be defined as an autonomous entity to pursue its own goals, which has the ability to perceive the environment, communicate with other agents, and take actions to affect the environment based on its knowledge and the results of communications.

In large-scale industrial systems various MAS technologies have been developed. A new approach based on the multi-agent system technology has been developed to cope with the dynamics and uncertainty of freight trains traffic management [2]. MAS technology applications to the modeling and simulation of urban public transportation networks has been proposed [3]. Novel agent-based architectures for manufacturing and evaluation are reported in [4][5].

The characteristic commonly existing in these problems is that the supervision or management functions of subsystems are naturally distributed. The task requires a large number of operators and various programs to work together to achieve the global goals. Each entity involved in the large-scale application has only a partial view of the entire scope, but the entire system can react in a way that all entities are closely coordinated [6].

Interactions and cooperation between agents are important features of MAS to achieve a global goal that is beyond the capability of each individual agent [7]. The performance of MAS can be determined by the interactions. Agents cooperate so that they can achieve more than they would if they act as individuals. Each agent, however, is autonomous and its decisions and actions are independent;

there is a possibility that these actions are in conflict. The coordination among agents can be achieved in an autonomous manner.

12.2 Strategic Power Infrastructure Defense System

A new defense system concept, Strategic Power Infrastructure Defense (SPID) system, has been developed by the Advanced Power Technologies (APT) Consortium consisting of the University of Washington, Arizona State University, Iowa State University and Virginia Polytechnic Institute. By incorporating multi-agent system technologies, the SPID system is able to assess power system vulnerability, monitor hidden failures of protective devices, and provide adaptive control actions to prevent catastrophic failures and cascading sequences of events.

12.2.1 SPID System Framework

The software agents of the SPID system for the power system vulnerability assessment and self-healing reconfiguration control are illustrated in Fig. 12.1. The architecture of the proposed multi-agent system is also shown in Fig. 12.1. The SPID system consists of three layers: reactive, coordination, and deliberative layers [8][9]. The agents in the reactive layer perform local subsystems or components control with a fast response time, while the agents in the deliberative layer analyze, monitor and control power systems from a global point of view. An important function of the middle (or coordination) layer agents is to examine the consistency of decisions received from the deliberative layer with the current

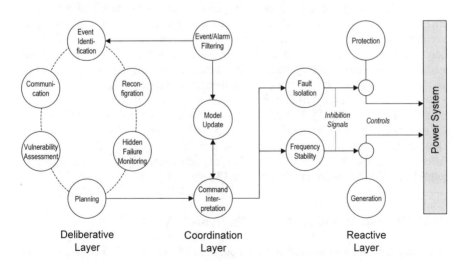

Fig. 12.1. Software agents in the SPID system

model of the power system. This is because the deliberative layer does not always reflect the current operating conditions of the power system and the decisions or plans from the deliberative layer might not be consistent with the current power system status. Another role of the coordination layer is to map the decisions from the deliberative layer into control signals that can be accepted by the agents in the reactive layer since the control decisions directly from the deliberative layer might be too condensed. The coordination between the deliberative and reactive layers is established on the "subsumption architecture" [17]. One advantage of the subsumption architecture is that local controls from the reactive layer can be inhibited by the deliberative layer if needed, based on the information at a global point of view. For example, due to a hidden failure of the phase-unbalance relays at the McNary power plant of the USA Western grid, all 13 generators were tripped in two minutes, which led the power grid to a catastrophic failure in August 1996 [18]. If the Western system were equipped with a control strategy that could analyze the impact of this hidden failure rapidly and inhibit the tripping signals in time, the 1996 outage could have been avoided or at least the impact could have been reduced. The decisions to inhibit the tripping signals can be made by the Vulnerability Assessment Agent in consideration of the system wide security after it analyzes the impact of tripping the generators.

The coordination of the software agents is an important issue since power system components and the software agents perform their tasks in a dynamic environment. The uncertainties in the power system require that the software agents have the abilities to adapt and learn. To achieve this goal, an intelligent and autonomous learning method is necessary for the agents to learn and update their capabilities through interactions with the environment. An important learning method is the supervised learning in which training data is provided to an agent by a supervisor within limited training time. This method has some limitations, for example, the software agent is passively guided through the task and it is often difficult for the supervisor to generate representative scenarios [10]. Another popular learning method is "reinforcement learning" that does not need to have an exact model of the environment and does not need a supervisor to learn. This method learns autonomously through interactions with the power system environment. If an action is followed by a satisfactory state of affairs, or by an improvement in the state of affairs, the tendency to produce the actions is strengthened (i.e. reinforced) [11]. For the agents' adaptive learning in the SPID system, the temporal difference method has been applied for the reinforcement function. The feasibility study of the reinforcement learning technique with the temporal difference method for an agent's adaptive learning capability with load shedding control schemes has been reported in [12][13].

12.2.2 Definitions and Roles of SPID System Agents

The definition and role of each software agent in the SPID system is given as follows [8][9].

Reactive Layer

Protection Agent: Each agent is modeled by protection logic, communication rules, and operation characteristics to represent a computer protective device such as a relay.

Generation Agent: Each agent represents a generating unit modeled by its MW and MVAR output capabilities, Automatic Generation Control (AGC) feature, excitation system, and other associated control systems.

Fault Isolation Agent: This agent identifies a fault or disturbance event by the protection scheme and transfers the diagnostic result to the Event / Alarm Filtering Agent. It has the ability to subsume the role of Protection Agents to prevent malfunctioning based on the information provided by higher layer agents.

Frequency Stability Agent: This agent determines frequency control actions such as generation control and load shedding. The agent can inhibit the actions of Generation and Protection Agents.

Coordination Layer

Event / Alarm Filtering Agent: This agent filters the input data to avoid excessive events / alarms. Only the triggering events / alarms that exceed threshold values are allowed to go through to the deliberative layer.

Model Update Agent: This agent continuously updates the model of the power system and it has the responsibility to check whether the decisions from the deliberative layer match the current system status.

Command Interpretation Agent: This agent decomposes the commands from the deliberative layer into control signals that will be transmitted to the reactive layer.

Deliberative Layer

Event Identification Agent: This agent identifies the outage portion of the network after it recognizes a fault or disturbance. This agent also generates the current power system model, which is shared by all the deliberative layer agents through the agents' communication channel.

Communication Agent: This agent provides fast and reliable network paths for critical control signals by continuously assessing communication system's status.

Hidden Failure Monitoring Agent: This agent monitors hidden failures of the power system components and reports to the Vulnerability Assessment Agent after it identifies the regions of vulnerability.

Vulnerability Assessment Agent: This agent continuously assesses the power system vulnerability to the sources of various threats including the information provided by the Hidden Failure Monitoring Agent.

Reconfiguration Agents: This model includes the self-healing algorithms that provide preventive and corrective control actions such as controlled islanding and load shedding when the system is in a vulnerable condition based on the result from the Vulnerability Assessment Agent.

Planning Agent: This agent determines the optimal sequence of control actions based on the self-healing plans provided by the Reconfiguration Agents in consideration of response time and resource constraints.

12.3 Controlled Islanding Agent of SPID System

As a multi-agent system, the SPID system is a distributed but coordinated network of problem-solving or decision-making agents working together to achieve a common goal that is beyond the capability of each individual agent [7]. The multi-agent technologies are proposed to accomplish complex tasks in a dynamic power system environment [9]. These complex tasks include assessing the vulnerability of the power grid with respect to possible cascaded events, and utilization of wide area protection and control that are responsive to the prevailing system conditions. In this section, new techniques are proposed for a "Controlled Islanding" Agent that serves as a reconfiguration agent of the SPID system.

12.3.1 Controlled Islanding of Power Systems

Power systems become more vulnerable as they are operated closer to their limits. Potential sources of vulnerability can be classified as internal and external sources to the power system infrastructure. Internal sources include power system component failures, protection and control system failures, information system and communication network failures, and inadequate security assessment. External sources are natural calamities, human errors in operations, gaming in electricity markets and sabotage [8]. Both kinds of sources may lead to catastrophic events and cascading outages. A final line of defense for unpredicted or unpredictable events are "adaptive protection, reconfiguration of the power network, controlled islanding, load shedding, and the ability of the system to survive the voltage and frequency excursions associated with the events" [8]. Among these defense schemes, the concept of controlled islanding has received considerable attention. The advantages of forced island formation for partial system preservation and system recovery have been shown in [14] by comparing with natural island formation. The islands formed by controls would be more stable and less prone to disintegration than those formed naturally. Moreover, system restoration will be less complex and the overall restoration time can be reduced if the island scheme is well designed and operated [14]. The successful operation of an Islanding Protection scheme prevented the customers of the metropolitan area in Tokyo from a blackout in 1999, in which an airplane crashed and damaged a 275 kV transmission tie line [15]. Various islanding schemes used for captive power plants in India are introduced in [16]. Without these islanding schemes, the severe disturbance of the power grid may result in significant loss in production, damages to process equipment, and even loss of life [16]. The above research results and industry operation experience serve to demonstrate that the controlled islanding scheme can play a significant role in the defense systems against catastrophic events.

Due to the restraints of network development and more intensive use of available generation and transmission facilities, a defense system is essential for reliable operation of interconnected power systems. A defense system is a protection scheme designed to identify the power system vulnerability that is known to cause unusual stress to the power system and to take predetermined preventive or corrective actions to reduce the vulnerability in a controlled manner. The goal of a defense system is to maintain the power system rotor stability, voltage stability, and acceptable power flows within the power network by taking appropriate control actions.

Controlled islanding is a last resort action for a large-scale power system following a severe disturbance that results in instability of the system. The traditional islanding strategies are mainly based on pre-specified operating conditions and fixed locations. In order to achieve the best possible security of the system, it is desirable to systematically build an islanding controller model into a wide-area monitoring, protection and control defense system framework such as the SPID system.

12.3.2 Context of SPID System Agents Interactions

Each agent in the SPID system attempts to achieve its individual goal and is relatively independent. The global goal is achieved by agents working together in the context of cooperative interactions. The lowest layer quickly reacts to perturbations while the highest layer evaluates the vulnerability of the system from a wide-area point of view. The control actions of the lowest layer can be modified or inhibited by a higher layer in order to obtain the coordinated control of the system. Thus, agents on different layers communicate through the layered structure in Fig. 12.1. Agents on the same layer also need to share their resources, information and decisions.

The flows of interactions among SPID system agents are shown in Fig. 12.2. Each agent is endowed with the communication capability. The communication between agents is at a knowledge level, which is similar to human interactions. The structure of the agents' interactions reflects the openness and flexibility of multi-agent systems.

It is noted that one of the Reconfiguration Agents of the SPID system is the Controlled Islanding Agent. By cooperative interactions, the Controlled Islanding Agent and other agents in Fig. 12.2 can provide preventive and corrective self-healing strategies to avoid catastrophic failures of a power system. When a Protection Agent detects a fault and decides to isolate the fault, the fault event will be transferred to the Fault Isolation Agent and at the same time to the Event/Alarm Filtering Agent. In order to avoid excessive events / alarms, only the triggering events / alarms that exceed threshold values are allowed to trigger the Event Identification Agent in the deliberative layer.

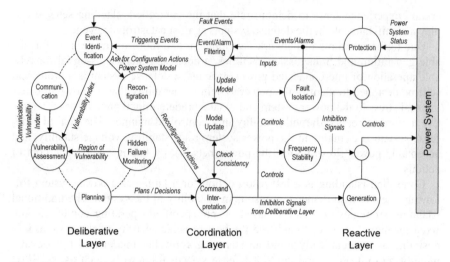

Fig. 12.2. Flows of interactions among SPID system agents

The Event Identification Agent will ask the Vulnerability Assessment Agent to assess the power system vulnerability due to the fault and the information provided by the Hidden Failure Monitoring Agent. If necessary, the Reconfiguration Agents, such as the Controlled Islanding Agent, will be required to provide self-healing actions to the Planning Agent where the optimal sequence of controls will be determined and sent out to the Command Interpretation Agent in the coordination layer. The Command Interpretation Agent first checks the consistency of the power system model of the plans with the Model Update Agent, which continuously updates the model of current power system by interactions with the Event / Alarm Filtering Agent. If the plans from the Planning Agent are out-of-date, the Command Interpretation Agent will ask the Event/Alarm Filtering Agent to trigger the deliberative layer agents to modify the plans. Otherwise the plans are valid and the Command Interpretation Agent will decompose them into actual control signals that will be transmitted to the reactive layer. These control signals may be inhibition signals that subsume the roles of Protection Agents and Generation Agents to prevent malfunctioning.

12.3.3 Controlled Islanding Agent

The SPID system is a multi-agent system that can be defined as a collection of heterogeneous, computational entities, each with its own problem solving capabilities and the ability to interact in order to reach an overall goal. The characteristics of an agent may vary based on application types. The following definition of an agent is based on the SPID system purpose:

- Which is capable of acting in an environment;
- Which can communicate directly with other agents;

- Which is driven by a set of tendencies in the form of individual objectives;
- Which possesses resources and knowledge of its own;
- Which is capable of perceiving its environment (to a limited extent);
- Which has only a partial representation of its environment.

Based on the above definition of a SPID system agent, the design of the Controlled Islanding Agent is illustrated in Fig. 12.3. The Controlled Islanding Agent consists of four components: information base, vulnerability assessment result acquisition, network and power flow data acquisition, and optimal islanding plan.

The Information base component connects with the communication network to exchange data with other SPID agents. Two popular agent communication protocols have been widely accepted for multi-agent system applications: FIPA (Foundation for Intelligent Physical Agents) and KQML (Knowledge Query and Manipulation Language) [19][20]. The SPID multi-agent system (SPIDMAS) uses FIPA for agent communication. The detailed descriptions of cooperative agent interactions using FIPA are provided in [9].

The vulnerability assessment result acquisition component extracts the current vulnerability assessment results from the information base, which continuously communicates with the Vulnerability Assessment Agent. This component will trigger the islanding plan component when it is necessary to island the system to avoid catastrophic events and cascading outages.

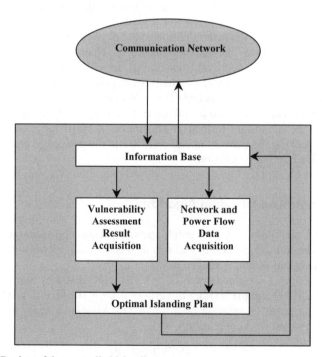

Fig. 12.3. Design of the controlled islanding agent

The network and power flow data acquisition component extracts the current power system network model and power flow data from the information base, which continuously exchanges the data with the Model Update Agent of the coordination layer.

After it receives the signal from the vulnerability assessment result acquisition component, the optimal islanding plan component starts to execute the islanding algorithm using the information from the network and power flow data acquisition component as its input. The output will be the optimal islanding plan that goes to the information base and then, through the communication network, to the Planning Agent. Whenever the Controlled Islanding Agent needs to send out a critical control signal, the Communication Agent provides the shortest communication paths to other agents.

12.3.4 Controlled Islanding Criteria

When the Vulnerability Assessment Agent indicates that the power system is approaching a catastrophic failure, Reconfiguration Agents such as Controlled Islanding Agent must take control actions to limit the extent of the damage. In our approach, the system is separated into smaller islands at a reduced capacity taking into account important aspects of restoration.

To implement the controlled islanding software agent, the determination of the islands, i.e., the points of separation, for a given operating condition is a critical step. A set of criteria for the determination of the physical boundary of each island is utilized. For example, the inherent structural characteristics of the system should be considered in identifying the islands. In addition, the choice of the islands should not be disturbance-dependent. An important consideration is the generation load imbalance in each island. The reduction of generation load imbalance in each island reduces the amount of under-frequency load shedding to be performed once the islands are formed. It also makes it easier for each island to be capable of matching the generation and load within the prescribed frequency limit and is beneficial during restoration.

12.3.5 A Controlled Islanding Algorithm

The islanding algorithm based on graph spectral partitioning techniques is being developed. The following theorem that sets up a connection between graph spectra and "Minimum Ratio Cuts" is proved in [21].

Theorem: Given a weighted graph $G = (V, E)$ with adjacency matrix A, diagonal degree matrix D and the number of vertices $|V| = n$, the second smallest eigenvalue λ_2 of the graph's Laplacian matrix $Q = D - A$ yields a lower bound on the cost c of the optimal ratio cut partition, with

$$c = e(U,W)/(|U| \cdot |W|) \geq (\lambda_2/n) \tag{12.1}$$

where V and E are the sets of vertices and edges of the graph G respectively, U and W are two disjoint sub-graphs after the partition, |U| and |W| represent the numbers of the vertices in the two sub-graphs, and $e(U,W)$ is the sum of the weights of the edges in $\{(u, w) \in E \mid u \in U \text{ and } w \in W\}$, i.e., the size of the cut set.

The result of the above theorem can be rewritten as

$$e(U,W) \geq (\lambda_2/n)\,(|U|\cdot|W|) \qquad (12.2)$$

Inequality (12.2) indicates that given a weighted graph G with n vertices and given a partition of disjoint U and W, the second smallest eigenvalue λ_2 of the graph's Laplacian matrix yields a lower bound on the size of the cut set. This suggests the following partitioning method: compute the second smallest eigenvalue λ_2 (Fiedler value) of the graph's Laplacian matrix Q(G) and the corresponding eigenvector x_2 (Fiedler vector), then use x_2 to construct the partition vector, from which the optimal partition can be found.

To construct the partition vector, the "median cut bisection" method and the general "m-partition" method have been proved to be an efficient way to identify the optimal cut set that minimizes the total weight of the cut-edges, i.e., the size of the cut set [22]. The "median cut bisection" method maps indices of vertices that have values above the median value of Fiedler vector x_2 to one part and below the median value of Fiedler vector x_2 to the other. The general "m-partition" method, namely a partition into m and (n-m) vertices, maps the m smallest components of x_2 to one part and the (n-m) largest to the other part.

Based on the graph-theoretic techniques discussed in [21][22], an efficient controlled islanding algorithm to separate a power system network into two islands (two-way partitioning) with the consideration of minimizing the generation load imbalance in each island is developed as follows:

- Compute the power flow of the power system network.
- Convert the power system network into a graph G.
  - Each bus of the power network is a node of the graph G.
  - Each transmission line of the one-line power system diagram is an edge of graph G.
  - The weight of each edge of graph G is assigned according to the absolute value of real power flow of the corresponding transmission line.
- Compute the Laplacian matrix Q of graph G.
  - Compute adjacency matrix A and diagonal degree matrix D of graph G.
  - The Laplacian matrix $Q = D - A$.
- Compute the second smallest eigenvalue λ_2 of Laplacian matrix Q.
- Compute x_2, the real eigenvector associated with λ_2.
- Map x_2 into a heuristic partition vector of the graph G.
  - Sort entries of x_2, yielding the sorted vector v of vertex indices.
  - Place all vertices in partition U.
  - For i = 1 to n-1
    - Move v_i from partition U to partition W.
    - Calculate the cut set size of the (U, W) partition.

- Find the optimal (U^*, W^*) partition that has the minimum cut set size among the above n-1 different partitions.

The optimal (U^*, W^*) partition will have the minimal generation and load imbalance on each island.

For illustration, the controlled islanding algorithm is applied to a small 6-bus power system as shown in Fig. 12.4. The weight of each edge is equal to the absolute value of real power flow of the corresponding transmission line. The second smallest eigenvalue λ_2 of the graph's Laplacian matrix is 0.1966 and its corresponding eigenvector is $x_2 = [0.4220\ 0.4218\ 0.3802\ -0.4101\ -0.4085\ -0.4055]$. By sorting the entries of x_2, the sorted vector v of vertex indices is obtained as $v =$ [Bus 4, Bus 5, Bus 6, Bus 3, Bus 2, Bus 1]. Table 12.1 lists the different n-1 partitions with corresponding cut set size. The optimal partition is obtained as island U = {Bus 4, Bus 5, Bus 6} and island W = {Bus 3, Bus 2, Bus 1} with the minimal cut size 0.3.

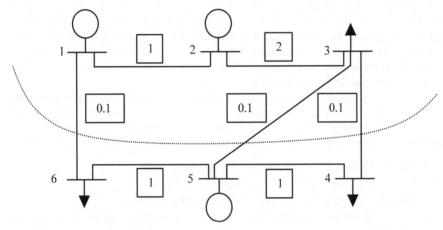

Fig. 12.4. The 6-bus power system network

Table 12.1. Partition results of the 6-bus power system

Partition {U, W}	Cut Set Size (Total weight of cut-edges)
{(4), (5 6 3 2 1)}	1.1
{(4 5), (6 3 2 1)}	1.2
{(4 5 6), (3 2 1)}	0.3
{(4 5 6 3), (2 1)}	2.1
{(4 5 6 3 2), (1)}	1.1

The above algorithm can be further expanded to achieve k-way partitioning when necessary, i.e., divide a power system network into k vertex disjoint islands. The approach is to recursively apply the spectral two-way algorithm to achieve the k-way partitioning. Furthermore, the reinforcement learning technique with the temporal difference method can be used to find the optimal sequence of islanding control actions.

12.4 An Application to MicroGrid Control and Operation

Distributed energy resources are attracting a great deal of attention because of their potential as important contributors in the power grid. The smaller generating units and renewable energy, e.g., micro turbines, wind and solar power, fuel cells and diesel generators, provide the resources that are critical at the peak load conditions. The availability of such generators also helps industrial, commercial or residential end-users avoid power outages caused by problems on the grid side. These distributed generators need to be connected to the grid in order to fully utilize their ability and increase the flexibility.

Traditional studies for distributed energy resources concentrated their efforts on the impact of these devices on the grid. However, most of them have been conducted for a small number of distributed energy resources. Moreover, "islanding detection", which ensures a secure shutdown of a connected distributed energy resource when a problem occurs on the grid side, is a major concern to avoid cascading failures. This is one of the typical subjects discussed in the past research.

This trend is natural if one assumes that a small number of distributed energy resources will be installed and connected directly to conventional distribution networks. However, based on this assumption, the capabilities of distributed resources are not fully taken advantage of.

Distributed energy resources can be installed close to loads. It is reasonable to keep the devices operating even when a problem occurs on the grid side. In addition, some of them have the capability to provide thermal load besides electricity, i.e., Combined Heat and Power (CHP). It is not practical to deliver the produced heat to sites far away due to the large thermal loss. Distributed energy resources and electrical and thermal loads should be located in a small area.

Technically, most of distributed energy resources are equipped with power electronic devices at their interfaces. These devices are flexible in control and their response can be very fast. A well-developed strategy can allow the distributed energy resources to demonstrate their full capabilities.

12.4.1 What is MicroGrid?

The term *MicroGrid* has several meanings in different fields. For example, micro-grid is the name of the simulator developed for Grid Computing performance evaluation in the computing / networking field.

However, MicroGrid here refers to a new concept to integrate various distributed energy resources and loads into a small grid. The concept has been proposed by several organizations, for example, the Consortium for Electric Reliability Technology Solutions (CERTS) [23], Electric Power Research Institute (EPRI), General Electric Co. etc. There are differences among them but the concepts are similar. In addition, there are other definitions or suggestions for this approach around the world. MicroGrid is identified by U.S. Department of Energy (DOE) as one of the critical research areas [24].

For the typical definition, reference [23] states:

"The Consortium for Electric Reliability Technology Solutions (CERTS) MicroGrid concept assumes an aggregation of loads and microsources operating as a single system providing both power and heat. The majority of the microsources must be power electronic based to provide the required flexibility to insure operation as a single aggregated system. This control flexibility allows the CERTS MicroGrid to present itself to the bulk power system as a single controlled unit that meets local needs for reliability and security."

MicroGrid has three distinguished features. First, MicroGrid provides an effective way to integrate distributed energy resources and loads in order to full utilize their abilities. MicroGrid can supply electrical and thermal loads simultaneously in effective ways even when a fault occurs on the system side. Secondly, MicroGrid can be a "good citizen" [23] or grid-friendly site to the surrounding distribution grid from the viewpoint of the utility system. This means that MicroGrid should be a self-controlled individual entity. Finally, MicroGrid provides a "plug-and-play" capability for each distributed energy resource, which meets the customers' local loads. Customers can install their own distributed energy resources as they like and it should not impact the grid. This is also an important feature to make MicroGrid behave as a "grid-friendly" entity.

Fig. 12.5 shows a typical implementation of MicroGrid. The MicroGrid structure is regarded as a single autonomous system supplying both electrical and thermal demands. In order to achieve high-level flexibility, the majority of distributed energy resources should be small generating devices with power electronic control.

12.4.2 MicroGrid Design

The example system proposed here, as shown in Fig. 12.5, has three feeders with electrical / thermal loads and distributed energy resources consisting of two microturbine generators, one fuel cell system, and a diesel engine driven generator. The MicroGrid is connected to the distribution grid through a Point of Common Coupling (PCC) located on the primary side of the transformer and defines the separation between the distribution grid and the MicroGrid. In the conventional research on distributed generators, the MicroGrid must satisfy the regulations or requirements at the connection point PCC.

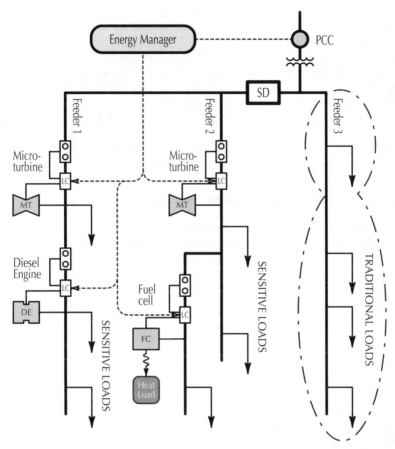

Fig. 12.5. Outline of the MicroGrid (PCC: Point of Common Coupling, SD: Separation Device)

Refer to ANSI / IEEE standard P1547 (Standard for Distributed Resources Interconnection with Electric Power Systems). This standard does not take into account the MicroGrid concept but it does allow for implementation of a group of distributed energy resources. To provide the MicroGrid with an adequate level of controllability, some regulations or requirements are similar to those of a large distributed energy resource.

Separation Device (SD) plays an important role in protection. There are two kinds of loads in this MicroGrid example, traditional loads and sensitive loads. Whenever a fault occurs or power supply from the grid is lost, sensitive loads must be isolated and met by distributed energy resources by opening SD as soon as possible. This is a major function of the MicroGrid called "islanding operation."

Each distributed energy source has its own Local Controller (LC) that is in charge of local control for the power flow, voltage and protection. This device belongs to the lowest layer of the multi-agent system architecture proposed here. The device must respond to disturbances or load changes as fast as possible without the need for communications.

The Energy Manager is the upper level controller that provides operational control through the dispatch of power generation and voltage set points to each Local Controller. Energy Manager can exchange necessary information with all Local Controllers through the communication network.

Flexible protection is also the important feature of the MicroGrid. Each distributed energy resource has a circuit breaker, not a sectionalizer, working with the Local Controller. Most distributed energy resources are interfaced with power electronic devices that provides the required capabilities.

Candidates for Distributed Energy Resources

Most of small-scale microturbines have a set of a rectifier and an inverter, which implements AC / DC / AC converter, since the turbine rotates at a highspeed and then generates high-frequency voltages. Fuel cells are DC generators based on chemical reactions. They also have inverters at their interfaces as a DC / AC converter. Other similar small generators can be employed to form the MicroGrid. The following is a list of the promising candidates to be considered:

- Microturbines,
- Fuel cells,
- Photovoltaic generators (PV),
- Wind turbines,
- Small hydro turbines,
- Storage systems,
- (Heat recovery systems),
- Reciprocating engines.

Existing reciprocating engines (e.g. diesel engine generators) are more familiar and competitive in price but they do not have power electronic devices. They cannot provide advanced and flexible control. This lack of flexibility can be compensated by other distributed energy resources such as microturbines.

12.4.3 MicroGrid Agent (MGA)

As mentioned previously, the Advanced Power Technologies (APT) Consortium has proposed a hybrid multi-agent system to implement the concept of a Strategic Power Infrastructure Defense (SPID) System [8]. It has been designed for the bulk power system but the concept itself can be adapted for the MicroGrid. Since the MicroGrid should be an autonomous and independent entity to the upper grid, and have to accommodate various smaller generators, it is natural to solve the grid control and operation problem by the multi-agent techniques. In addition, the conventional control center is not economically feasible for the MicroGrid. Once the multi-agent system methodology is developed, the system can be further enhanced with learning and adaptation capabilities.

In this study, an intelligent multi-agent system is proposed that is able to handle the technical and economic functions of a "control center" for a MicroGrid. If each generating device and load has the ability and intelligence to communicate with the MicroGrid control agent and work together to solve technical and economic problems, then the MicroGrid can be managed by a multi-agent software system. The proposed MGA must be able to identify the technical and economic problems to be mentioned in the following subsections.

The concept of the MGA is shown in Fig. 12.6, which has *subsumption architecture* [17]. This structure is based on SPID but there are some differences in the roles of some agents since SPID system was designed as a defense system. MGA should have ability to handle both of control and operation problems but does not have to be a high-level defense system. The different features are discussed in the following

Reactive Layer

Generation Agent: Each Generation Agent represents a distributed energy resource characterized by the kW and kVAR outputs and capabilities, as well as the essential features of load sharing capabilities. Most distributed energy resources are equipped with power electronic devices. The generation agent is modeled by operating parameters and control modules.

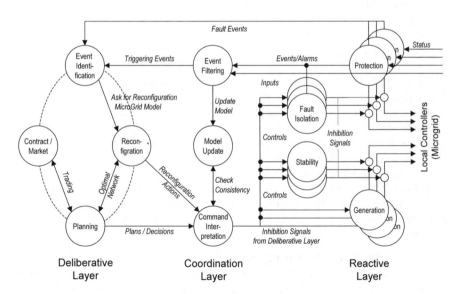

Fig. 12.6. MicroGrid Agents (MGA)

Protection Agent: Each Protection Agent represents a protective device (e.g. relay or SD). Following a fault, the agent can take autonomous actions to protect components in a MicroGrid. The protection agent is modeled by relay logic, rules, operating characteristics, and communications with other agents. For the Micro-Grid, SC is a special protective device that should respond faster than the other protective devices and needs coordination with them.

Stability Agent: This agent determines the frequency control actions, including generation control and load shedding to maintain the MicroGrid stability. This agent must recognize the operating condition, islanding or connecting. This agent can inhibit actions of the generation agent.

Fault Isolation Agent: MicroGrid should take care of various kinds of loads, e.g. traditional or sensitive loads. The required security level depends on the importance of each load. When a fault occurs, Protection Agents react immediately. However, the protection scheme should be determined from a global point of view. This agent can provide inhibition signals to Protection Agents according to the commands provided from the upper layer.

Coordination Layer

Model Update Agent: This agent updates the current real-world model and checks whether the plans (commands) from the deliberative layer are consistent with the current status of the system. For the MicroGrid, there are various types of distributed energy resources. Each of them can connect to or disconnect from the MicroGrid. This agent has to pay attention to the information provided by Generation Agent.

Deliberative Layer

Planning Agent: This agent finds the optimal switching sequence among feasible sequences identify the plans and decisions made by other agents in the deliberative layer. The plans, including switching sequences, are transmitted to the command interpretation agent. For the MicroGrid, this agent also determines the operational strategy including economic dispatch and electricity trading. Planning Agent and Contract / Market Agent can cooperate to tackle this issue.

Contract / Market Agent: The SPID system does not include this agent, which determines if an operation plan violates the contract. The MicroGrid is connected to the grid under the contract agreed upon by both sides. The plan prepared by the Planning Agent should be checked by this agent to make the MicroGrid a grid-friendly site.

Fig. 12.7 illustrates an implementation of the MGA corresponding to the example system of Fig. 12.5. The *Reactive Layer* agents are allocated to Local Controllers and the other upper layer agents are gathered to the Energy Manager site.

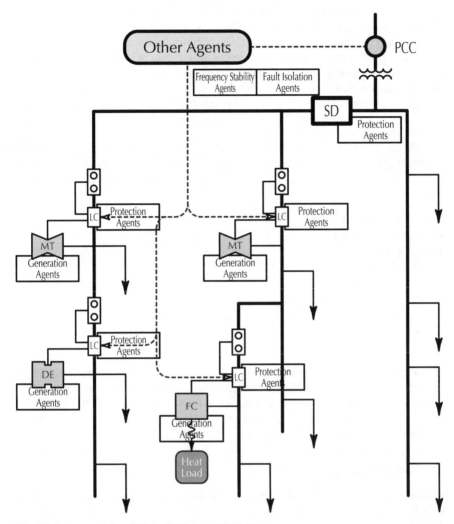

Fig. 12.7. An example of MicroGrid Agent (MGA)

12.4.4 MicroGrid Control

As mentioned in the previous section, a MicroGrid can include various distributed energy resources and loads with interface to the grid. This interface can increase reliability. However, in order to operate the MicroGrid as an isolated system, the MicroGrid must have the ability to regulate the frequency, voltages, and be able to maintain grid security when it is not interconnected to the grid. The MicroGrid may also participate in trading of electricity in the market.

Traditionally, the frequency and voltage control of a power grid is performed by the automatic generation control (AGC) system and the voltage control systems including shunt capacitors, generator reactive power outputs, and transformer tap changers. The conventional control center has a number of system operators, power traders and other technical and business support staffs. The control center is also equipped with the energy management systems with the computer hardware, software and communication facilities linking the control centers to various sub-stations in the power system. For the MicroGrid, it is not economically feasible to provide such a sophisticated control center.

This issue is dealt with by Local Controllers, which should work together with other Local Controllers or supervised by upper-level controller of Energy Manager under the operation conditions. The major control issues to be considered are as follows:

- Real and reactive power control
- Voltage regulation
- Load tracking
- Load sharing
- Frequency regulation
- Grid protection against faults
- Grid security assessment, including system dynamics
- Power quality
- Trading of electricity

Note that the technical problem is non-trivial. The control of distributed energy resources is different from the traditional fossil or hydro units. For example, the control of wind turbines is slower and microturbines and fuel cells have different characteristics comparing with the conventional synchronous generators. Some of them will be briefly discussed in the following subsections.

Real and Reactive Power Control

Distributed energy sources like diesel-driven synchronous generators can be controlled in the traditional way. However, most of distributed energy sources are equipped with a voltage source inverter at their interfaces since they have DC sources or high-frequency generators to be rectified. The distributed energy sources are connected to the MicroGrid through inductors as shown in Fig. 12.8.

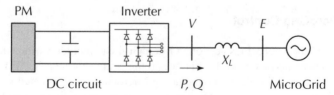

Fig. 12.8. Inverter equipped distributed energy resources

Real power P and reactive power Q from the distributed energy resource to the MicroGrid through the interconnected inductor reactance X_L are given by a set of well-known equations:

$$P = \frac{V \cdot E}{X_L} \sin \delta_p \tag{12.3}$$

$$Q = \frac{V(V - E \cos \delta_p)}{X_L} \tag{12.4}$$

$$\delta_p = \delta_v - \delta_e \tag{12.5}$$

Where, V: inverter output voltage magnitude, E: bus voltage magnitude, X_L: connection inductor, δ_v: phase angle of V, δ_e: phase angle of E.

Relationship between power P and phase difference δ_p is nonlinear. However, for the small value for δ_p, it can be regarded as almost linear since $\sin \delta_p \approx \delta_p$. Similarly, Q is predominantly dependent on V. Since the voltage source inverter can control the magnitude of the output voltage V and its phase angle δ_v simultaneously, these relationships are used to implement a feedback loop for the control of output power P and voltage V.

Generation Agents are in charge of this control. The set point of the output is controlled by a supervised signal provided from the deliberative layer (Planning Agent). If necessary, control signals provided from the Generation Agents are inhibited by the Stability Agents.

Voltage Regulation

In the bulk power grid, impedances between conventional generation units are large enough to restrict circulatory phenomena of reactive currents. However, MicroGrid may not have a large central generation unit; instead, it consists of a number of distributed energy resources. Basic real and reactive power control mentioned above is not enough to regulate the voltage for local stability. If the Local Controller does not have the voltage regulation control, there is a possibility for voltage or reactive power oscillations to take place among the distributed energy resources. Local Controllers must have the capability to restrict large circulatory phenomena of reactive currents.

Fig. 12.9 shows the function of a voltage-reactive current control for Local Controllers. When the generated reactive current becomes capacitive, the set point of local voltage for Local Controller is reduced, and vice versa. Of course, there is Q-limit value Q_{max} in this droop characteristic as a function of VA ratings of its inverter and / or the power supplied from the prime mover.

Load Sharing and Frequency Regulation

One of the main features of the MicroGrid is its islanding operation. The MicroGrid can provide flexible power similar to an Uninterrupted Power Supply (UPS) even if it is disconnected from the grid.

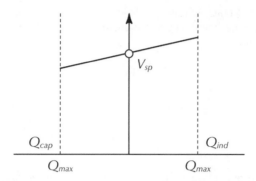

Fig. 12.9. Voltage regulation

This means the MicroGrid has adequate controllability to achieve a smooth transition for islanding and reconnecting autonomously. In the island operation mode, distributed energy resources should share the total loads existing in the MicroGrid because no power is supplied from the grid. Similar to the conventional power system, the power-frequency droop characteristic at each Local Controller should achieve the goal without the need for communications.

Fig. 12.10 shows an example of the power-frequency droop function. Two distributed energy resources are operating at P_1 and P_2. When the MicroGrid enters into the islanding operation mode, the frequency drops from the nominal value f_0 to the new equilibrium point f_1 to share the total load. Then, the Local Controller proceeds to the frequency regulation process to restore the frequency to the nominal value f_0 while maintaining the output power level.

On the other hand, when connected to the grid, the loads in the MicroGrid can be supplied both from the grid and from the distributed energy sources. This is a normal operation and the set point of power output of distributed energy sources are controlled by the Local Controllers according to the operation signals provided by the higher level controller, Energy Manager. This is an analogy of the conventional control center working for the bulk power systems.

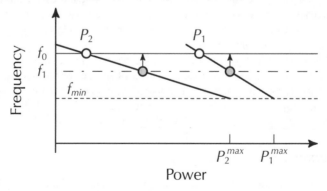

Fig. 12.10. Load sharing (frequency droop)

All Local Controllers must have the capability to quickly respond to the emergency power sharing without communication, and to communicate with Energy Manager or other Local Controllers for the necessary information exchange under the normal operation.

Generation Agents can contribute to this type of control. Emergency controls, such as immediate load sharing, have to be taken as soon as possible without communication. However, under the normal operation, each distributed energy resources should be operated according to the plan managed by the deliberative layer. The plan can be optimized in several senses, e.g., economical, environmental and so on.

Load Tracking

Bulk power systems have many generators that store energy in their rotating masses. The inertia is large and its kinetic energy can contribute to the energy balance immediately when a load changes, which results in a small drop of system frequency.

However, there may not be a central, large generation unit inside the MicroGrid and most distributed energy resources have small or no inertia. As mentioned above, the MicroGrid should be implemented to achieve a smooth transition from grid-connected mode to the islanding mode, and vice versa. Under the islanding mode, load-tracking problems may appear since the MicroGrid has various kinds of energy sources with different characteristics.

12.4.5 MicroGrid Operation Issue

A MicroGrid can have various distributed energy sources. Their economic characteristics are different from each other as well as their dynamics. Some of them are suitable for base load operation and the others may be good at responding to sudden load changes. In addition, several types of them have the capability to supply thermal load besides electricity. By fully utilizing their capabilities, the MicroGrid can be operated efficiently and economically.

There is an analogy of Best Mix Strategy and Economic Load Dispatch (ELD) for the bulk power systems except CHP functions. Moreover, if a contract and a regulation permit, the MicroGrid can sell their electricity to the market through the grid. This is an incentive for investment in the MicroGrid.

The economic consideration of the MicroGrid is a major issue since there are a number of small distributed energy sources with different characteristics and the interaction with the grid should be taken into account for selling / buying electricity from / to the utility.

This mode of operation is usually planned and carried out at the central control center for the bulk power system. For the MicroGrid, however, it is not practical to provide a dedicated control center, and Energy Manager will be in charge of this task. The multi-agent system can be a promising option to achieve the autonomous Energy Manager that can serve some critical functions of the conventional control centers.

In the MGA, agents in deliberative layer can give a good solution, in particular, Planning Agent and Contract / Market Agent. The Planning Agent prepares an operation plan and sends a message to the Contract / Market Agent. The agent checks the plan according to the concluded contracts and the current situation of the electricity market. If the plan violates the contracts, two agents can negotiate to solve the conflicts. Then, the Planning Agent can send the plan to the Command Interpretation Agent. Each Generation Agent holds the information about its own generating unit. The Model Update Agent always updates the real-world model through the communication with each Generation Agents. When the plan is submitted to the Command Interpretation Agent, the Model Update Agent can check the plan whether the plan is surely based on the correct model.

12.4.6 An Example for Oscillation Restriction

One of the most important features of the MicroGrid is the islanding operation and plug-and-play capability. When the MicroGrid enters to the islanding operation mode, load sharing and frequency regulation problem will arise as a major concern. In the MicroGrid, there are various distributed energy resources and there may not be a large generating unit. The conventional reciprocating engine generator has a slow response for load tracking and the other inverter equipped energy sources have quick responses. The response is quite different among various distributed energy sources.

Load Sharing under the Islanding Operation

An example of the oscillation restriction capability of the MGA to achieve the good performance of islanding operation is discussed. Microturbines and diesel-driven synchronous generators are typical examples of distributed energy resources.

Most of small microturbine generators have a set of a rectifier and an inverter as an AC / DC / AC converter since the turbine rotates at high speed and generates high-frequency voltages. The response of the inverter is very fast but the turbine and generator have small inertia in their rotating masses and it induces time delay in response. On the other hand, diesel generators have are synchronous generators, which are connected to the MicroGrid directly without inverters. This leads to some difficulty in its response to the load change. When the MicroGrid is connected to the grid, the load change can be compensated by the grid and the frequency regulation is also achieved. On the other hand, distributed energy sources inside the MicroGrid are responsible for load sharing and frequency regulation in the islanding mode.

In this example, an oscillation between two types of different generators is shown and a brief introduction of countermeasure to restrict the oscillation by using MGA is given.

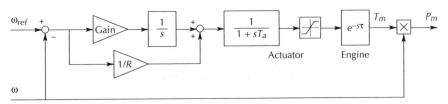

Fig. 12.11. Diesel engine model

Diesel Engine Generator Model

Reciprocating engine model has been well developed and the authors adopted one of the typical implementations, for example in [26][27]. Fig. 12.11 shows a block diagram of the governor model, including dynamics of the engine and its controller. One of the key points of the behavior is the time delay induced from the crankshaft dynamics.

The exciter model is the IEEE Type 1 model [28] with appropriate parameters and the generator part can be modeled as a conventional synchronous generator model with a small inertia. In this example, the second order Park model was applied. They are well known and not shown here.

Microturbine Generator Model

Several types of microturbine generation systems are available in the market. The most popular type has a high-speed turbine, which is combined with a permanent magnetic exciter generator and a set of AC / DC / AC converter. The prime mover dynamic model has been developed and examined in the literature, for example in [29]. Fig. 12.12 shows the dynamics model of the microturbine, which includes governor, turbine and the controller behavior. The produced high-frequency AC power is rectified to DC, and then converted to AC power with nominal frequency 60 Hz by using the voltage source inverter.

Oscillation Problem

The test system is shown in Fig. 12.13. This system has two diesel-driven synchronous generators, two microturbine generators and loads. In the normal operation, assume that the total load is larger than the total generation by four distributed energy resources. That means the loads are supplied from the four distributed energy resources and the adjacent grid. If the grid power is lost for some reason, all loads should be shared by distributed energy resources only.

This simulation has been implemented and executed on Matlab / SIMULINK [30]. The grid is modeled as an infinite bus and the distribution lines within the MicroGrid are modeled as a π section equivalent.

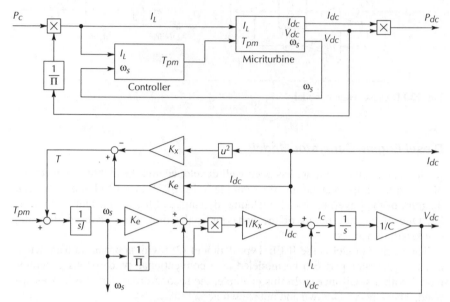

Fig. 12.12. Microturbine dynamics model

The response of generators is shown in Fig. 12.14. At the beginning, the MicroGrid is connected to the grid, and then disconnected at 3.0 sec. Power sharing is conducted based on the power-frequency droop function. The frequency also returns to the nominal value. However, an oscillation phenomenon was observed in the output power of the microturbine, and the frequency restoration process is slow. It seems that the oscillation is induced from the difference of response speed between the two types of generators. The diesel-driven synchronous generators are connected directly but the microturbine generators are connected through the inverters. Inverters control the voltage angle difference and can respond very quickly.

Oscillation Restriction by MGA

In order to avoid the undesirable responses, Generation Agents and Stability Agents (with MGA) can contribute to achieve the oscillation control. Generation Agents have the information about its own distributed energy resources, for example, generator type, control systems, and their parameters. When the distributed energy resource is connected to the MicroGrid, its Generation Agent can send the information to the Model Update Agent and / or other agents to inform that a new distributed energy resource has participated. The control parameters are determined by upper-layer MGA according to the MicroGrid structure and operation strategy.

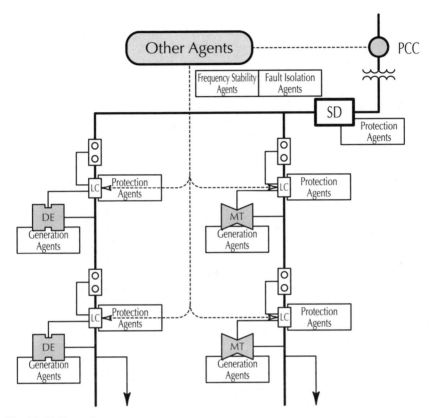

Fig. 12.13. Example system

An example of Generation Agents for the MGA is shown in Fig. 12.15. This consists of three parts, System Model, Diagnosis and Decision / Communication part. The System Model part is in charge of holding its own model and control system parameters. Diagnosis plays a role of self-checking. If there is a technical problem on the unit, this part can identify the problems. Decision / Communication part communicates with other agents to successfully adapt itself to the changing environment, and sends the control command to the Local Controllers.

Event Identification Agent in the deliberative layer is constantly monitoring the events and acquiring the data. This agent identifies the MicroGrid situation and generates the current MicroGrid model. When Decision / Communication part perceives that the distributed energy resource is connected to the MicroGrid, it sends a message to the Model Update Agent including the information stored in the System Model part. The Event Identification Agent and the Model Update Agent can execute several typical simulations to produce the adaptive control plan. The produced plan is checked through the other agent in the deliberative layer and sent to the Command Interpretation Agent.

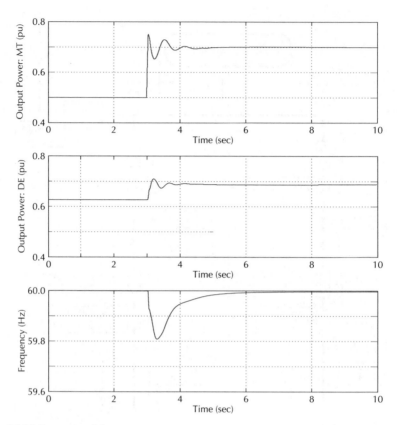

Fig. 12.14. Response of the test system

Then, the control command is sent to the Generation Agent and the Stability Agent. In this case, the Generation Agent responds and sends a control signal to the Local Controller but the signal may be inhibited by the Stability Agent according to the adaptive control plan.

Adaptive learning method is an important issue for the application of the multi-agent systems. The method proposed for the MGA is the Reinforcement Learning technique, same as that of the SPID system [12][13].

In this section, an application of multi-agent system for the MicroGrid, called MGA, is proposed. Since the MicroGrid has a number of distributed energy resources and loads and it must be autonomous, the multi-agent system concept is suitable for intelligent control and operation of the MicroGrid. The MGA can be an autonomous control system, which is very different from the conventional control center for bulk power systems. This section also provides a discussion based on the power oscillation problem due to different kinds of distributed energy resources in the islanding mode. One of the key features of the MicroGrid is its islanding operation mode and therefore the undesirable phenomena should be improved.

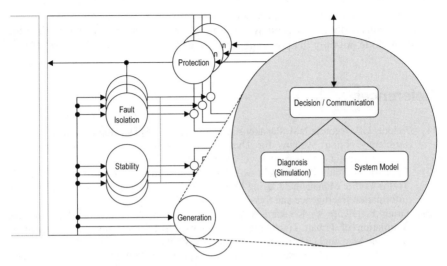

Fig. 12.15. An example of Generation Agents for the MGA

Moreover, another key function of the MicroGrid is the "plug-and-play" capability that allows various types of distributed energy resources to be connected to the MicroGrid. The characteristics or response times are quite different. Hence, the MGA should determine the best control strategy each time they are connected.

12.5 Conclusions

In this chapter, the applications of multi-agent system technologies to the controlled islanding for the SPID system and the MicroGrid control are discussed.

Controlled islanding is generally a last resort control for a large-scale power system following a severe disturbance that results in instability of the system. The conventional islanding strategies are mainly based on pre-specified operating conditions and fixed locations. By implementing the Controlled Islanding Agent as a wide-area protection and control tool of the SPID system, the vulnerability of the grid can be reduced. By incorporating multi-agent system technologies, the SPID system is able to assess power system vulnerability, monitor hidden failures of protective devices, and provide adaptive control actions to prevent catastrophic failures and cascading sequences of events. The new measurement technologies, for example, the phasor measurement unit (PMU) based wide-area measurement network built on the top of a powerful LAN/WAN, can guarantee the real-time delivery of synchronous phasors and control signals [31].

For the MicroGrid, an application of multi-agent system is proposed as MGA (MicroGrid Agent). MGA was designed to control various kinds of distributed energy sources according to the characteristics of the MicroGrid. The MicroGrid also has the potential to be a cost-effective concept. This function can be achieved by utilizing the upper-level agents of MGA. However, this is more complicated than the conventional situation for the bulk power systems since the planning should include not only the power generation but also the supply of heat (CHP).

References

[1]   Lekkas GP, Avouris NM, Papakonstantinou GK, (1995) Development of Distributed Problem Solving Systems for Dynamic Environments. IEEE Trans. on Power Systems, Man, and Cybernetics, vol 25, pp 400-414

[2]   Cuppari A, Guida PL, Martelli M, Mascardi V, Zini F (1999) Prototyping Freight Trains Traffic Management Using Multi-Agent Systems. International Conference on Information Intelligence and Systems, pp 646-653

[3]   Gruer P, Hilaire V, Koukarn A (2001) Multi-Agent Approach to Modelling and Simulation of Urban Transportation Systems. IEEE International Conference on Systems, Man, and Cybernetics, vol 4, pp 2499-2504

[4]   Jarvis D, Jarvis J, McFarlane D, Lucas A, Ronnquist R (2001) Implementing a Multi-Agent Systems Approach to Collaborative Autonomous Manufacturing Operations. Aerospace Conference, IEEE Proceedings., vol 6 , pp 2803-2811

[5]   Brennan RW, William O (2000) A Simulation Test-Bed to Evaluate Multi-Agent Control of Manufacturing Systems. Simulation Conference Proceedings, vol 2, pp 1747-1756

[6]   Ferber J (1999) Multi-Agent Systems: An Introduction to Distributed Artificial Intelligence. Addison-Wesley

[7]   Grossman RL, Nerode A, Ravn AP, Rischel H (1993) Hybrid Systems. Springer-Verlag

[8]   Liu CC, Jung J, Heydt GT, Vittal V, Phadke AG (2000) Conceptual Design of the Strategic Power Infrastructure Defense (SPID) System. IEEE Control System Magazine, Aug., pp 40-52

[9]   Jung J, Liu CC (2001) Multi-Agent Technology for Vulnerability Assessment and Control. Proceedings. of IEEE Summer Meeting, pp 1287-1292

[10]   Sutton RS, Barto AG, Williams RJ (1992) Re-Inforcement Learning in Direct Adaptive Optimal Control. IEEE Control System Magazine, April, pp 19-22

[11]   Sutton R, Barto AG (1998) Reinforcement Learning: An Introduction. MIT Press

[12]   Jung J, Liu CC, Tanimoto SL, Vittal V (2002) Adaptation in Load Shedding Under Vulnerable Operating Conditions. IEEE Trans. on Power Systems, vol 17, Issue 4, pp 1199-1205

[13]   You H, Vittal V, Jung J, Liu CC, Amin M, Adapa R (2002) An Intelligent Adaptive Load Shedding Scheme. 14th Power Systems Computation Conference (PSCC), Sevilla, Spain

[14]   Archer BA, Davies JB (2002) System Islanding Considerations for Improving Power System Restoration at Manitoba Hydro. IEEE Canadian Conference on Electrical and Computer Engineering (CCECE 2002), vol 1, pp 60-65

[15]   Agematsu S, Imai S, Tsukui R, Watanabe H, Nakamura T, Matsushima T (2001) Islanding Protection System with Active and Reactive Power Balancing Control for Tokyo Metropolitan Power System and Actual Operational Experiences. Seventh International Conference on (IEE) Developments in Power System Protection, pp 351-354

[16] Rajamani K, Hambarde UK (1999) Islanding and Load Shedding Schemes for Captive Power Plants. IEEE Trans. on Power Delivery, vol 14, no 3, pp 805-809

[17] Brooks RA (1991) Intelligent without Reason. Massachusetts Institute of Technology, Artificial Intelligence Journal (47), pp 139-159

[18] (1996) WSCC Disturbance Report for the Power System Outage that Occurred on the Western Interconnection, [Online] www.wscc.com

[19] Labrou Y, Finin T (2003) A Proposal for a New KQML Specification. University of Maryland. [Online] www.cs.umbc.edu/kqml

[20] Chiariglione L (1998) FIPA 98 Specification, Foundation for Intelligent Physical Agent. [Online] www.cselt.it/fipa/spec/fipa98

[21] Hagen L, Kahng AB (1992) New Spectral Methods for Ratio Cut Partitioning and Clustering. IEEE Trans. on Computer-Aided Design of Integrated Circuits and Systems, vol 11, pp 1074-1085

[22] Chan TF, Ciarlet Jr. P, Szeto WK (1997) On the Optimality of the Median Cut Spectral Bisection Graph Partitioning Method. SIAM Journal on Computing, vol 18, no 3, pp 943-948

[23] Lasseter R, Akhil A, Marnay C, Stephens J, Dagle J, Guttromson R, Melio-poulous AS, Yinger R, Eto J (2002) Integration of Distributed Energy Resources: The CERTS MicroGrid Concept. White paper prepared for U. S. Department of Energy, California Energy Commission

[24] Office of Power Technologies (2001) Energy Efficiency and Renewable Energy, Department of Energy. Transmission Reliability Multi-Year Program Plan FY 2001-2005

[25] Jenkins N, Allan R, Crossley P, Kirschen D, Strabac G (2000) Embedded Generation. IEE, London

[26] Roy S, Malik OP, Hope GS (1991) An Adaptive Control Scheme for Speed Control of Diesel Driven Power-Plants. IEEE Trans. on Energy Conversion, vol 6, no 4, pp 605-611

[27] Stavrakakis GS, Kariniotaks GN (1995) A General Simulation Algorithm for The Accurate Assessment of Isolated Diesel–Wind Turbines Systems Interaction, Part I: A General Multimachine Power System Model. IEEE Trans. on Energy Conversion, vol 10, no 3, pp 577-583

[28] IEEE Committee Report (1973) Dynamic models for Steam and Hydro Turbines. IEEE Trans. on Power Apparatus and Systems, PAS-92, pp 1906-1915

[29] Lasseter RH (1998) Control of Distributed Resources. Bulk Power System Dynamics and Control IV – Restructuring, Santorini, Greece

[30] Mathworks Co. [Online] http://www.mathworks.com/

[31] Kamwa I, Grondin R, Hebert Y (2001) Wide-Area Measurement Based Stabilizing Control of Large Power Systems – A Decentralized/Hierarchical Approach. IEEE Trans. on Power Systems, vol 16, no 1, pp 136-153

13 Operation of Quality Control Center Based on Multi-Agent Technology

Hiroyuki Kita, Jun Hasegawa

Division of System and Information Engineering, Graduate School of Engineering
Hokkaido University, Sapporo, Japan

Flexible, Reliable and Intelligent ENergy Delivery System (FRIENDS) is a new framework of future electric power systems proposed by the authors. In FRIENDS, new facilities called Quality Control Center (QCC) are installed into the distribution systems, and the electricity with some kinds of quality can be produced in QCC corresponding to the customers' requirements. This section explains a method for the emergency operation and the voltage regulation in the distribution systems characterized by FRIENDS. The proposed methods are constructed based on the multi-agent technology using bilateral communication between QCCs; therefore, autonomous and distributed operation can be realized.

13.1 Introduction

Recently, interests in distributed generators (DGs) such as fuel cell, micro gas turbine, photo voltaic cell, and secondary battery are growing because of their higher efficiency and lower impact on the global environment. On the other hand, introduction of a number of DGs into electric power distribution systems involves the following issues, which are quite different from those in the conventional distribution systems:

- Difficulties in voltage regulation by the occurrence of reverse power flows from DGs,
- Increase in short circuit current,
- Difficulties in fault detection and clearance.

"Flexible, Reliable and Intelligent ENergy Delivery System (FRIENDS)"[1] is a new future distribution system proposed by the authors, which can resolve not only the above problems concerning introduction of DGs, but also several potential problems under the deregulated environment of electric power industry. The

most original and important idea of FRIENDS is to install some new facilities called *"Quality Control Centers (QCCs)"* between distribution substations and customers as shown in Fig. 13.1. The QCC includes power electronics devices for improving power quality, and DG and battery for efficient utilization of the electric energy or as backup sources at the fault occurrence. Therefore, every QCC can be considered just as a small sized electric power plant with quick response characteristics. Further, computers for processing information are also installed in QCC, and all of them are interconnected to each other through the information communication network. This chapter presents a method for operating a couple of QCCs dispersively and autonomously by applying the multi-agent technology. In the proposed method, every QCC is modeled as an agent that can behave autonomously by data collected locally and information obtained through communications with the neighboring agents. As a result, more quick and exact operation of the distribution system can be realized.

13.2 Multi-Agent and FRIENDS

An agent is a component of the multi-agent system, which is able to sense a state of an environment and affect the environment according to its own autonomous decision-making. The basic idea of the multi-agent system is to realize more complex behaviors as a whole by direct or indirect interactions among agents. Further, it has advantages of higher fault tolerance and faster behavior than the conventional centralized system.

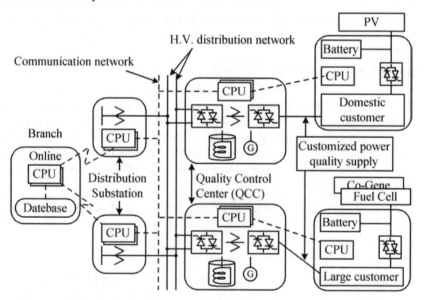

Fig. 13.1. General image of FRIENDS

The quality control center (QCC) in FRIENDS can affect the topology of the distribution network and the supply reliability and quality of the electricity to customers directly or indirectly by controlling the interior devices and communicating among the other QCCs. Therefore, the concept of the multi-agent system can be applied to the distributed and autonomous operation in FRIENDS by assigning QCCs and distributed substations as agents and the other system components such as feeders or customers as the environment, respectively. First, this chapter applies the multi-agent system to an emergency operation in FRIENDS, which can be applied as a temporary operation to be implemented immediately after a fault occurrence. Next, it proposes a method for regulating the distribution voltages in FRIENDS autonomously.

13.3 Agent Models

13.3.1 Model of Quality Control Center

Fig. 13.2 shows a model of the interior structure of QCC. Hybrid Transfer Switch (HTS) which consists of a mechanical switch (MS) and anti-paralleled thyristor switches (TS) is equipped on each feeder at the high voltage side (22kV, resistance grounding system) in QCC [2]. In order to reduce the conductive losses, the line current in the distribution network flows only through MS when HTS is closed, while TS is used only when HTS received a signal for turning on or turning off.

This section uses the UPS-type QCC proposed by the authors as a model of the low voltage side (415/240V, solid grounding system) in QCC. This model consists of two PWM controlled inverters (Inv. 1 and Inv. 2), DG and solid-state switches. Controlling these devices adequately makes it possible to produce the electricity with three kinds of quality called ordinary quality (OQ), high quality (HQ) and super premium quality (SP) defined in Table 13.1 [3].

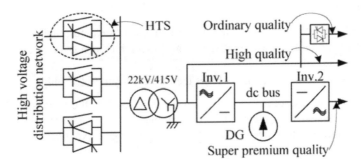

Fig. 13.2. The model of a QCC.

Table 13.1. Definition of Power Quality

	waveform	reliability
OQ	not improved	low
HQ	not improved	high
SP	improved	very high

More specifically, the OQ and HQ lines are connected to the distribution network directly through transformer; therefore, the distortion of waveform quality in the distribution network cannot be improved. On the other hand, in the load that the SP power is supplied, the voltage is provided by Inverter 2. Therefore, the load is not affected by the distortions of waveform quality in the distribution network because it is separated by the DC bus. Further, the QCC can continue to supply the electricity even in the faulted state by using DG in QCC as a backup source. More specifically, when QCC is separated from the distribution network by opening HTS, the loads with OQ and HQ are supplied by the DG through Inverter 1 and the load with SP is supplied through Inverter 2. Here, if capacity of DG is less than total power demand in QCC, the load with OQ is cut off first according to the predefined priority.

13.3.2 Model of Distribution Substation

Distribution substation (DSS) is also modeled as an agent. Figure.13.3 represents an interior structure of DSS in the FRIENDS. Some switches for turning on or turning off the feeders connected to the neighbor QCC are equipped in DSS.

13.3.3 Information Measured

In the concept of the multi-agent technology, each agent measures only local states in the environment to reduce the amount of data processed by each agent. Therefore, this section assumes that each agent can measure the following information.

- On/off state of the high voltage feeder that agents are connected to
- Active and reactive powers and current flowing into agents along each feeder.
- Receiving voltage.
- Load demand of each quality level fed through agent (QCC only).

The data on the other agents is not actively used in the proposed algorithm; therefore, data-sharing mechanism such as blackboard system is not needed. The data required are informed by some kinds of message as explained in the next section.

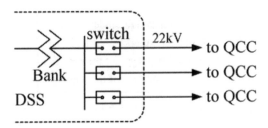

Fig. 13.3. The model of distribution substation

13.3.4 Communication between Agents

Communication between agents in the proposed operation is similar to the message passing of ABCL/1 [4]. Each agent can communicate only with the neighbor agents using a data packet called "message" in this section. Here, the neighbor agent means an agent connected by electric power distribution line and optical fiver cable for the communication of messages directly and physically. Therefore, the neighbor agents of each agent are immovable in both the normal and the faulted situation and each agent is able to find the message-passing route.

Message consists of sender (From), destination (To), type of message (MessageType) and contents (Contents) as shown in Fig. 13.4.

13.4 Emergency Operation of Distribution Systems

13.4.1 Operational Policies

In order to realize the distributed and autonomous emergency operation, the six types of message shown in Table.13.2 are used. The emergency operation proposed in this section is based on the following four policies.

1. Clearance of the faulted line is given priority to every operation.
2. Each agent supplies the electricity to its own customer prior to the other agent's one.
3. Each agent is willing to give the active and reactive powers, if it is asked to help from the neighbor agent. If there is not enough reserve of DG to supply the requested power, the agent gives as much powers as possible.
4. Each agent asks the neighbors to give the power only when it does not have enough capacity of DG.

> From : ID of message sender
> To : ID of destination
> MessageType : Type of the message
> Contents : Contents of the message

Fig. 13.4. Component of message

Table 13.2. Types of message

MessageType	Explanation
FLT (FauLT)	Announcement of the fault detection. Content is ID number of the faulted line.
SA (Stand Alone)	Order for the neighbors to enter stand alone operation
SRQ (Supply ReQuest)	Request for passing power. Contents are required active and reactive power, the original sender of SRQ and priority
SP (Supply Possible)	Reply for SRQ (if possible to pass). Contents is the original SRQ sender
CSR (Cancel SRq)	Cancellation of SRQ. Content is the original sender of SRQ

The policy (1.) is needed to improve the supply reliability of the whole distribution network. The other three policies are based on the assumptions that each agent gives priority to supplying electricity to its own customers and help the neighbor QCCs if possible. These assumptions (especially, policy (2.)) would be valid if all QCCs are owned only by one entity. However, when some QCCs belong to the different entity from the other QCCs, these assumptions may be invalid. In order to simplify the problem, this section supposes that every agent is generous to the cooperative emergency operation.

13.4.2 Proposed Algorithm

The action of each agent is triggered by message arrivals. Each agent decides its own action autonomously considering data collected locally and information obtained through the communications with the neighbors as explained in the following subsections.

Action for FLT

FLT is generated only when a fault detector installed in each agent has detected the fault occurrence. How an agent behaves after it has received FLT can be categorized into the following three cases depending on the location of the agent.

- Case 1: the agent is not connected to the faulted line electrically: The agent does nothing because it does not have the information about the topology between the agent and the faulted line.

- Case 2: the agent is connected to the faulted line and separated from DSS by clearing the faulted line: Since the agent cannot be supplied the electricity from DSS after clearing the faulted line, it initiates a stand-alone operation by the following steps.

  - Send message SA to the neighbors.
  - Turn off every line connected to the agent (including the faulted line) and supply electricity to the customers by DG.
  - If capacity of DG is insufficient, cut off the load with OQ and send message SRQ to the neighbors.

- Case 3: the agent is connected to the faulted line, however, it can be supplied from DSS even if the faulted line is cleared: First, the agent turns off the faulted line. Then, DG and Inverter 1 in the agent generates the active and reactive powers as much as it has been fed through the faulted line. By this action, the post-fault power flows in the distribution system are maintained to the same power flows as pre-fault (Fig. 13.5); therefore, it can avoid the fault extensions due to line overloading.

Action for SA

After the agent has received SA, it behaves according to the same steps as Case 2 of the previous subsection. By this action, every QCC, which was separated from DSS by clearing the faulted line, initiates a stand-alone operation.

Action for SRQ

If the capacity of DG is enough to provide the requested power, the QCC changes the output of DG and replies SP to the agent, which sent SRQ. Further, the distribution line connecting both of the agents is charged by turning on the HTS. On the other hand, if the capacity of DG is insufficient and impossible to provide the requested power, the QCC propagates SRQ to the other neighbors to obtain the shortage power. This "SRQ propagation" constitutes a cooperative group in which the requested power is transmitted via multiple agents, and this manner of cooperation is called "cascade" in this chapter.

In the FRIENDS network, loop configuration is permitted. Therefore, plural SRQs that an agent sent may be arrived to the identical agent through the different routes. In order to solve this conflict, the information about "priority" is attached to SRQ. SRQ with a higher priority is adopted if an agent received plural SRQs derived from the identical agent.

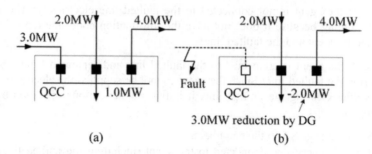

Fig. 13.5. Maintaining the power flows

More specifically, if an agent received SRQ with a lower priority than the one, which has been already received, SRQ which received later is ignored (Fig. 13.6 (a)). On the other hand, if an agent received SRQ with a higher priority than the one, which has been already received, the SRQ, which received earlier, is overwritten by the later SRQ (Fig. 13.6 (b)).

Further, there is also a possibility that an agent receives plural SRQ generated by the different agents. In such situation, the QCC deals with the SRQ according to the order of arrival. More specifically, the QCC stacks the received SRQ in FIFO database named "SRQ list" and does not cope with the next SRQ until it receives messages CSR or THA for the current SRQ.

When DSS received SRQ, it replies SP to the agent which sent SRQ sender if the capacity of transformers and switches are larger than the requested power.

Action for SP

If the agent, which received SP, is the one that sent SRQ, it replies THA to the agent that sent SP (and turns on switches connected to the line if it has not been turned on yet). And also, it sends CSR to the other neighbors in order to nullify the SRQ that has been already sent.

On the other hand, if the agent, which received SP, is not the one that sent SRQ, it propagates SP to the previous agent along the cascade route and sends CSR to the other neighbors.

Action for CSR and THA

The agent, which received CSR, propagates CSR to the neighbors in order to nullify the SRQ that has been already sent. Further, output of DG is set to its original state if that agent has already begun generating the requested power.

The agent that received THA propagates THA along the cascade route if it is not a terminal of cascade. Further, if the corresponding data can be found in SRQ list, it is removed from the list and the agent initiates to deal with the next SRQ, if its SRQ list is not empty.

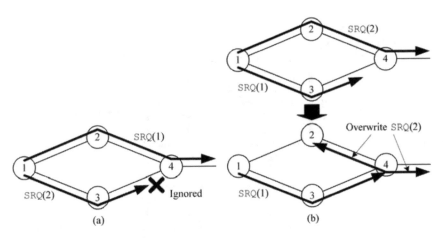

Fig. 13.6. Priority of SRQ

13.4.3 Simulation Results

In this section, the performance of the proposed method was analyzed through the Monte Carlo simulations. The twenty-five QCCs system illustrated in Fig. 13.7 was used as a test system and the fault occurrence at the distribution line between QCC21 and DSS3 was assumed. As an index for evaluating the performance of the emergency operation, the duration time of interruption (ID) was employed.

In order to obtain an exact distribution of ID, the 1000 simulations were executed. Turn off delay of HTS was given by a uniform random number from 2.0 to 3.0 msec when no fault current flows, and from 6.7 to 10.0 msec when the fault current exists [2]. Further, the demand of each QCC was set by a uniform random number from 0.9 to 1.1 MW.

Effect of DG Capacity on ID Distribution

In order to find the relationship between the capacity of DG installed and the ID distribution, the Monte Carlo simulation was carried out. In the simulation the following three cases were assumed as the capacity of DG.

- Case 1: 25 - 35% of total demand
- Case 2: 45 - 55% of total demand
- Case 3: 65 - 75% of total demand

More specifically, when a QCC is separated from DSS, it has to cut off its OQ load and send SRQ to the neighbors for a shortage in the capacity of DG. Fig. 13.8 shows the obtained ID distribution of QCC12, 16, 21 (message delay of 1.0 - 2.0 msec (uniform random) is applied).

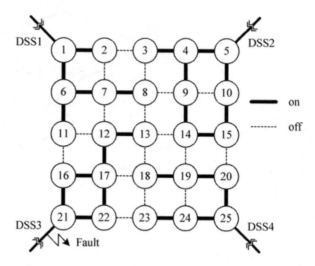

Fig. 13.7. Twenty-five QCCs and four DSSs model system for evaluating the performance quantitatively and maintaining the power flows

As shown in Fig. 13.8, ID was longest in QCC 21 and was shortest in QCC16. This trend of ID distribution is based on the following reasons:

- Since the emergency operation of each QCC is triggered by the arrival of the message, the QCCs, which are far from the faulted line, initiate the emergency operation later than the QCCs, which are close to the faulted line.
- Since the plural SRQs are coped with according to the order of arrival, the above delay yields longer delay time.
- The QCCs, which are far from the QCCs, being operated in the normal condition have to form the cascade via more QCCs being operated in the faulted state.

The simulation result also indicates that the increase in DG capacity improves (shortens) the ID distribution and it is more effective for the QCC that has the worse ID distribution.

Effect of Message Delay on ID Distribution

Effect of message delay on ID distribution was also investigated through the Monte Carlo simulation. The simulation was carried out with the following three message delays:

- Case 1: 1.0 - 2.0 msec
- Case 2: 2.0 - 3.0 msec
- Case 3: 3.0 - 4.0 msec

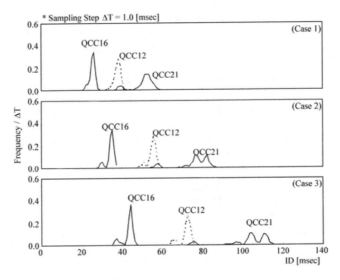

Fig. 13.8. Communication delay and ID distribution

Capacity of DG wais set to 25 - 35% of total demand; therefore, the cooperation with the other QCCs is needed to satisfy all demands. Fig. 13.9 shows ID distributions of QCC12, 16 and 21 obtained through the simulation. The simulation result implies that ID becomes longer as the message delay becomes longer and the relationship between them is almost linear.

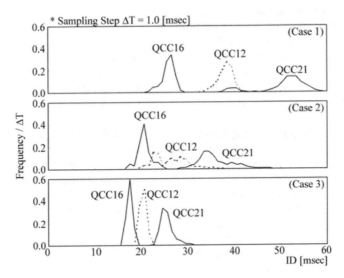

Fig. 13.9. DG capacity and ID distribution

The reason of this linearity is as follows. The number of messages required for completing the emergency operation does not depend on the message delay, however it depends on the final network configuration formed by the proposed methods. Further, the final configuration depends on the capacity of DG as shown in the previous result.

From the above two simulation results, we can find that increase in capacity of DG can level IDs and use of the fast information communication network can improve ID of the whole network. Further, the above simulation results indicate that the proposed method can realize the fast (tens of msec order) emergency operation if DGs with adequate capacity are installed and the infrastructure that can communicate the messages quickly is used.

Here, the response delay of DG was not considered in the above simulations. If the response time of DG is longer than the IDs in Fig. 13.8 and 13.9, it becomes a bottleneck and elongates the ID distributions. However, DG is only one backup generator in each QCC, therefore, it also has to supply the electricity to the SP load without any voltage waveform distortions and fluctuations. From this reason, the installation of DG with fast response time or energy storage system with fast response time such as micro-SMES or secondary battery in parallel to the DG is needed. Therefore, this chapter assumes to be negligible short as the response delay of DG.

13.5 Voltage Regulation of Distribution Systems in FRIENDS

As a method for regulating the distribution voltage in FRIENDS, there would be two kinds of approach. One is a centralized control approach in which the optimal operation of every QCC can be determined while collecting and monitoring information about the whole power system. The other is an autonomous and distributed control approach in which every QCC determines the most optimal operation on its own only using local information around it. This research presents a hybrid voltage regulation method in which the above two approaches are combined appropriately.

1. Autonomous Voltage and Reactive Power Control: In QCC, some facilities are installed for improving the power quality. By absorbing and generating the reactive power using these facilities the distribution voltage can be regulated adequately. Here, taking account into the local characteristics of the distribution voltage, the autonomous and distributed approaches can be used

2. Reconfiguration in Network Topology: This research proposes a method for reconfigurating the network topology by static transfer switches, which are installed in QCCs to regulate the distribution voltage. In this case, every QCC cannot estimate the post-reconfiguration condition of power system only using information around it because a bit of change in the system topology affects the states of the whole power system. Therefore, this chapter uses the centralized

control approach to determine the optimal network topology. The overview of the hybrid operation proposed in this research can be shown in Fig. 13.10.

13.5.1 Autonomous Voltage and Reactive Power Control in FRIENDS

The voltage and reactive power control in each QCC is composed of two kinds of task. In the first task every QCC absorbs or generates the reactive power corresponding to changes in the own demand by controlling some apparatus inside QCC. This is called Q-TBC because it is similar to the Tie-line Bias frequency Control (TBC) among multiple areas. In another task every QCC changes its reference value of the voltage magnitude within its permitted limits based on the information communication among QCCs. It can be realized by the concept of the multi-agent system. Both of the above can be explained in the following sections in detail.

Application of Q-TBC to Voltage Regulation in QCC

In Q-TBC method in reference [5], the whole transmission system can be divided into some sub-systems. Each sub-system behaves autonomously so that the internal balance between supply and demand in the reactive power is satisfied. The change in the reactive power demand inside the sub-system can be estimated by the voltage magnitudes at the own buses connecting to the other sub-systems and reactive power flows in tie-lines between subsystems. The change in the reactive power demand can be vanished by operating some reactive power sources inside the own sub-system.

In FRIENDS, the Q-TBC method can be applied similarly by assigning QCC as the above sub-system and considering the resistive elements in the distribution lines, which can be neglected in the transmission lines. The reactive power to be controlled in QCC can be calculated as follows.

First, giving the reference value of the voltage magnitude at QCCi and QCCj as V_{i0} and V_{j0} as shown in Fig. 13.11, the reference value of the reactive power at the middle point of tie-line between both QCCs can be calculated as follows:

$$Q_{ji0} = \frac{V_i^2 \left(V_{j0}^2 - V_{i0}^2 \right) - R_{ij} \left\{ 2P_{jiinj} V_i^2 + R_{ij} \left(P_{jiinj}^2 + Q_{jiinj}^2 \right) \right\}}{2 X_{ij} V_i^2} \qquad (13.1)$$

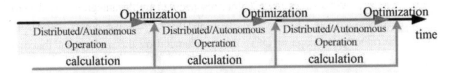

Fig. 13.10. Distributed and autonomous operation and centralized operation

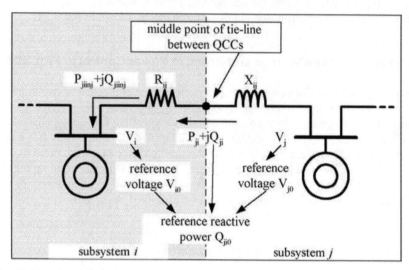

Fig. 13.11. Subsystem for Q-TBC

Here, the sign of Q_{ji0} is positive when the reactive power is flowing from QCCj to QCCi.

V_i : voltage magnitude measured at QCCi, P_{jiinj} and Q_{jiinj} :real and reactive powers injected to QCCi measured in the tie-line between QCCi and QCCj, R_{ij} and X_{ij} are resistance and reactance of tie-line between QCCi and QCCj.

The reactive power at the middle point of the tie-line can be calculated using P_{jiinj}, Q_{jiinj} and V_i which can be measured easily:

$$Q_{ji} = Q_{jiinj} + \frac{X_{ij}\left(P_{jiinj}^2 + Q_{jiinj}^2\right)}{2V_i^2} \tag{13.2}$$

Using Q_{ji} in Eq. 13.2, the reactive power to be compensated in QCCi, Q_{AR_i} can be calculated as follows:

$$Q_{AR_i} = \sum_j \left(Q_{ji} - Q_{ji0}\right) + K_i \left(V_i - V_{i0}\right) \tag{13.3}$$

Here, K_i is a constant value that relates the voltage magnitude to the reactive power in QCCi. To maintain the voltage magnitude of QCCi at the reference value, some reactive power sources in QCCi are rescheduled by the integral control until Q_{AR_i} becomes zero. If Q_{AR_i} exceeds the capacity of reactive power sources in QCCi, the reference voltage of QCCi can be changed by the multi-agent function explained in the following section.

Cooperative Voltage Regulation in FRIENDS by a Multi-Agent System

In FRIENDS, every QCC can communicate with the other QCCs by receiving or sending some kinds of information through the information communication network. This research applies the concept of the multi-agent system so that multiple QCCs can cooperate with the voltage and reactive power control. The proposed algorithm considers two types of message as shown in Table.13.4. The whole algorithm of the voltage and reactive power control in this research is shown in Fig. 13.12. Every QCC always monitors operating conditions of the own QCC and checks whether or not the messages from the other QCCs were arrived. If any of the following patters are satisfied, then the QCC behaves adequately and autonomously.

- Pattern 1: When Q_{AR_i} obtained using the current reference value of the voltage magnitude of QCCi exceeds the capacity limit of its reactive power source, the QCCi updates the reference value so that the violation can be alleviated. Here, if the updated voltage magnitude exceeds the permitted value, go to the next condition.
- Pattern 2: QCCi sends the message (DEV) for asking the neighboring QCC to update the reference value of the voltage magnitude so that the Q_{AR_i} can be alleviated.

Table 13.3. Two kinds of message for VQ control

MessageType	Explanation
DEV	Request the change in the reference voltage
VCH	Announce of the change in the reference voltage

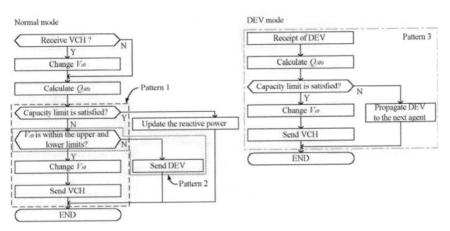

Fig. 13.12. VQ control algorithm

- <u>Pattern 3:</u> When QCCj receives DEV from QCCi, QCCj calculates the reactive power to be rescheduled by updating the reference and checks whether or not it exceeds the capacity of the reactive power source in QCCj. If the capacity limit is satisfied, the reference value of voltage magnitude V_{j0} is updated. Otherwise, QCCj sends the message to the next neighboring QCC to ask the cooperation more widely. Here, in every condition, when V_{i0} was updated, it sends the message VCH to all of the neighboring QCCs to inform the change in the reference value.

Simulation Results

The effectiveness of the proposed algorithm based on the Q-TBC and multi-agent system was ascertained through numerical simulations using a simple model system as shown in Fig. 13.13. In this model system, real and reactive power demands in every QCC are 1.0 pu and 0.1 pu respectively. The capacity of the reactive power source inside QCC is 0.5 pu. At the time of 40 msec after the simulation is started, real and reactive power demands in QCC6 were changed up to 1.6 pu and 0.16 pu respectively. The simulation was executed every 0.01 msec and the time required for communicating between QCCs was assumed to be 1.5 msec. Also, the reactive power sources in QCC can reschedule the reactive power in proportion to 1.0 pu/msec. Fig. 13.14 shows the reference and actual values of voltage magnitude in every QCC, and the rescheduled reactive power in each QCC. From Fig. 13.14, QCC6 increases the reactive power to maintain the voltage magnitude at its reference value immediately after the load increase. However, after the reactive power source attained its capacity, QCC6 started updating the reference value of voltage magnitude, V_{60}. Further, when V_{60} attained its lower limit, it sent the message, DEV to QCC3. When QCC3 could not cooperate with QCC6 any more, DEV was sent to the next neighboring QCC. By such cooperation, the voltage magnitude of QCC6 was restored within its lower and upper limits. Here, note that the way for controlling the reactive power in QCC4 is different from that in the other QCCs. Since QCC4 is far from QCC6, which caused the load change, QCC1 and QCC5 received the message DEV from QCC6 before QCC4. Therefore, the two QCCs have raised their reference voltage magnitudes first. As a result, Q_{AR_i} of QCC4 was alleviated.

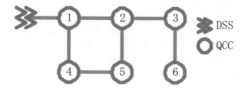

Fig. 13.13. Model system for simulation.

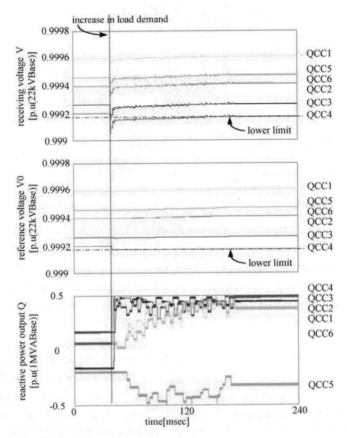

Fig. 13.14. Simulation result (VQ control)

13.5.2 Reconfiguration of QCC Network Topology in FRIENDS

There is a case that the proposed autonomous voltage and reactive power control method cannot be implemented by the shortage in the capacity of reactive power source in QCC. Also, when one QCC violates its upper limit and the other QCC violates its lower limit, the voltage magnitude cannot be maintained only by the proposed method. In this section, a method for changing the topology of QCC network is used with the autonomous voltage and reactive power control.

The problem for obtaining the optimal topology of QCC network can be formulated as a problem for determining tap setting of the distribution substation and open/close of switch in every distribution line. The objective function to be minimized is defined as follows:

$$f(\mathbf{x}) = \sum_{QCC_i} \left(V_{center,i} - V_i(\mathbf{x}) \right)^2 \tag{13.4}$$

The constrained conditions are defined as follows:

$$\text{line limits:} \quad I_i(\mathbf{x}) \leq \overline{I}_i (i \in line) \tag{13.5}$$

$$\text{voltage constrains:} \quad \underline{V}_i \leq V_i \leq \overline{V}_i (i \in QCC) \tag{13.6}$$

$$\text{short circuit capacity constraints:} \quad S_i(\mathbf{x}) \leq \overline{S}_i (i \in QCC) \tag{13.7}$$

$$\text{loss constraints:} \quad \sum_i LOSS_i(\mathbf{x}) \leq \overline{LOSS}(i \in line) \tag{13.8}$$

network constraints: every QCC is connected only to one substation $\quad (13.9)$

Here, $\overline{V}_i$ and $\underline{V}_i$ are upper and lower limits of V_i. $V_{center,i}$ is a center value of V_i which can be calculated by $V_{center,i} = (\overline{V}_i + \underline{V}_i)/2$. $\mathbf{x}$ is a vector composed of tap settings of distribution substation and 0-1 variables corresponding to open/close of switch in every distribution line. I_i is the current in line i. $\overline{I}_i$ is the upper limit of I_i. S_i is the short circuit capacity at QCCi. $\overline{S}_i$ is the upper limit of S_i. $LOSS_i$ is the distribution loss in line i. $\overline{LOSS}$ is the upper limit of loss in the whole distribution system.

As shown in Eq. 13.4, the objective function of the proposed method is defined as the sum of a square of deviation in the voltage magnitude from $V_{center,i}$. Spreading the margin of the voltage in every QCC makes possible to implement the voltage and reactive power control more easily. Also, even if the distribution loss is considered as the other objective function, the hybrid operation of this chapter can be applied similarly.

This chapter applies the Genetic Algorithm (GA) for solving this problem. The model system, which composed of two distribution substations and twenty-five QCCs as shown in Fig. 13.15 was used for the simulation. Two kinds of load were set in every QCC. One is a heavy load and the other is a light load as shown in Table 13.4. In the latter case, there are reverse flows in some QCCs. Fig. 13.16 shows the obtained network topology in each case. In the heavy load, the topology with many loops was obtained so that the voltage drop becomes smaller as a whole. On the other hand, in the light load the radial topology was obtained so that some QCCs which results in the reverse flow was located in the end point of the system. This topology contributes to alleviate the voltage drop. Here, the population size of GA in this chapter was set to 100 and the generation size was set to 500. The computation time was about 8 min. with a personal computer of Pentium III (1GHz).

Table 13.4. Load data

Active power of load [p.u.] (p.f.=0.9)									
Heavy load condition									
QCC	P	QCC	P	QCC	P	QCC	P	QCC	P
1	1.0	6	1.0	11	1.0	16	1.0	21	1.0
2	1.0	7	1.0	12	1.0	17	1.0	22	1.0
3	1.0	8	1.0	13	1.0	18	1.0	23	1.0
4	1.0	9	1.0	14	1.0	19	1.0	24	1.0
5	1.0	10	1.0	15	1.0	20	1.0	25	1.0
Light load condition (with reversal power flow)									
QCC	P	QCC	P	QCC	P	QCC	P	QCC	P
1	0.2	6	0.3	11	0.4	16	0.5	21	0.3
2	0.2	7	0.4	12	-0.3	17	0.5	22	0.2
3	0.3	8	0.5	13	-0.5	18	-0.5	23	0.3
4	0.2	9	-0.3	14	0.4	19	0.2	24	0.3
5	-0.2	10	0.5	15	0.3	20	-0.2	25	0.2

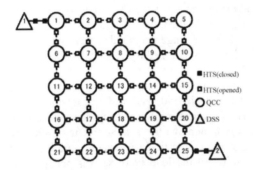

Fig. 13.15. Model system for simulation

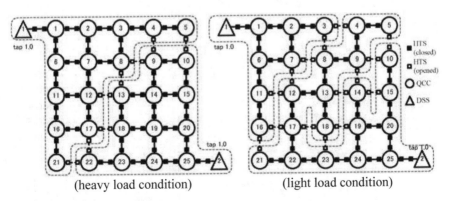

(heavy load condition)   (light load condition)

Fig. 13.16. Simulation results

13.5.3 Evaluation of the Performance of the Voltage Regulation

In a hybrid method proposed in this chapter, an autonomous voltage and reactive power control is implemented together with the reconfiguration of the QCC network topology. To verify the effectiveness of the proposed method, this section simulated the voltage regulation using three kinds of method; (1) autonomous voltage and reactive power control, (2) reconfiguration of the QCC network topology and (3) the hybrid method with both control. The purpose of those methods is to regulate the voltage in the distribution systems; therefore, a sum of a square of deviation of the voltage magnitude from its reference value was calculated over a time period for every QCC and its mean value per one QCC was used as a performance index.

The simulation was implemented using the model system in Fig. 13.15. The reactive power sources with the capacity of 0.5 MVar are installed in every QCC. The assumed time period is 30 min and the reconfiguration was implemented every 10 min considering the computation time evaluated in section 5.2. The load data of every QCC was set using the random number so that the peak value becomes around 1.0 MW. The load factor was set as 0.9 (lag) in every QCC. The real power demand used is shown in Fig. 13.17. First, the configuration obtained at every 10 minutes is shown in Fig. 13.18. The voltage magnitude at every QCC is shown in Fig. 13.19 and Fig. 13.20 for three kinds of method. Also, the value of the performance index is shown in Table 13.5.

From the comparison between Fig. 13.19 (a) and Fig. 13.20, fluctuation of the voltage magnitude by the hybrid method became smaller than that only by the voltage and reactive power control. Also, by comparing between Fig. 13.19 (b) and Fig. 13.20, it was shown that the voltage fluctuation became smaller by reconfiguration of QCC network. Particularly, reconfigurating at the time of 1200 sec when every QCC initiated the load increase was pretty effective for the voltage regulation. The above conclusion can be ascertained even by performance index in Table 13.5.

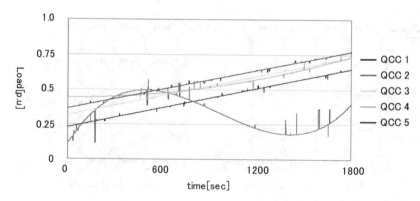

Fig. 13.17. Load data

Table 13.5. Voltage Index

Only Topology Reconfiguration	1.55537e-6
Distributed and Autonomous Operation	1.87322e-8
Proposed Operation (Hybrid)	1.85784e-8

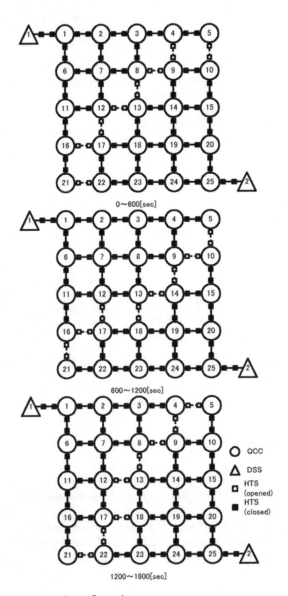

Fig. 13.18. Optimal network configuration

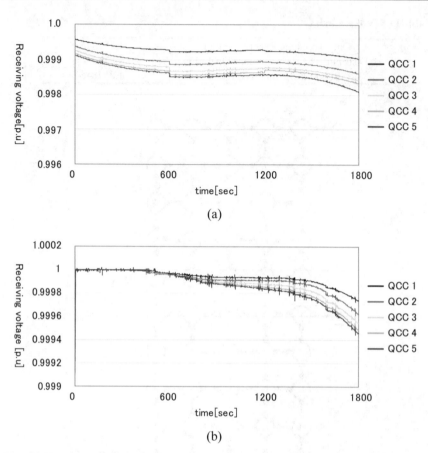

Fig. 13.19. Simulation results. (a) Only topology reconfiguration. (b) Distributed and autonomous operation.

13.6 Conclusion

This chapter proposed a multi-agent based emergency operation method and voltage regulation in FRIENDS. In the proposed methods, each QCC and DSS is modeled as an agent and each agent decides its behavior in distributed and autonomous manner. The effectiveness of the proposed method was also verified through the numerical simulations. From the simulation results, it was confirmed that more quick and exact operation can be implemented by the proposed method.

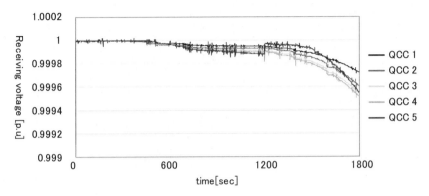

Fig. 13.20. Simulation results (hybrid operation)

References

[1] Nara K, Hasegawa J (1997) A New Flexible, Reliable and Intelligent Electrical Energy Delivery System. Trans. on Electrical Engineering in Japan, vol 121, no 1, pp 26-34

[2] Hara R, et al. (2000) The effect of the Transfer Switching on the Low Voltage Side of Quality Control Center in FRIENDS. Proc. of the 2000 International Conference on Power System Technology (POWERCON2000), vol. II, pp 667-671

[3] Hara R, et al. (1999) Proposal of Interior Structure and Operation of Quality Control Center in Flexible, Reliable and Intelligent Electrical Energy Delivery System. Proc. of International Conference on Electrical Engineering 1999 (ICEE'99), vol 2

[4] Yonezawa A (1990) ABCL: An Object-oriented Concurrent System", MIT Press

[5] Tanimoto M, et al (2000) An Autonomous Distributed VQC Method based on Q-TBC. Trans. of the IEEJ, vol 120-B, no 6, pp 815-822

Index

Druck: Strauss Offsetdruck, Mörlenbach
Verarbeitung: Schäffer, Grünstadt